Climate Change and Environmental Hazards Related to Shipping

Climate Change and Environmental Hazards Related to Shipping

An International Legal Framework

Edited by

Hans-Joachim Koch
Doris König
Joachim Sanden
Roda Verheyen

Proceedings of the Hamburg International Environmental Law Conference 2011

MARTINUS NIJHOFF PUBLISHERS

LEIDEN • BOSTON
2013

Library of Congress Cataloging-in-Publication Data

Hamburg International Environmental Law Conference (2011)
Climate change and environmental hazards related to shipping : an international legal framework : proceedings of the Hamburg International Environmental Law Conference 2011 / Edited by Hans-Joachim Koch … [et al.].
p. cm.
ISBN 978-90-04-24392-7 (hardback : alk. paper)—ISBN 978-90-04-24495-5 (e-book)
1. Water—Pollution—Law and legislation—Congresses. 2. Climatic changes—Law and legislation—Congresses. 3. Maritime law—Congresses. 4. Law of the sea—Congresses. 5. Marine pollution—Law and legislation—Congresses. 6. Marine resources conservation—Law and legislation—Congresses. I. Koch, Hans-Joachim, 1944– II. Title.

K3585.8.A6H36 2013
344.04'6343—dc23

2012041225

This publication has been typeset in the multilingual "Brill" typeface. With over 5,100 characters covering Latin, IPA, Greek, and Cyrillic, this typeface is especially suitable for use in the humanities. For more information, please see www.brill.com/brill-typeface.

ISBN 978-90-04-24392-7 (hardback)
ISBN 978-90-04-24495-5 (e-book)

This book is printed on acid-free paper.

FSC
www.fsc.org
MIX
Paper from responsible sources
FSC® C004472

PRINTED BY DRUKKERIJ WILCO B.V. - AMERSFOORT, THE NETHERLANDS

CONTENTS

PART ONE

WELCOMING ADDRESSES

PART TWO

CLIMATE CHANGE

Presentations

EU and National Initiatives

PART THREE

ENVIRONMENTAL HAZARDS RELATED TO SHIPPING

Presentations

PART SIX

CITIES' CONTRIBUTIONS TO ENVIRONMENTAL PROTECTION

PART SEVEN

RESULTS

FOREWORD BY THE EDITORS

As 'European Green Capital 2011', Hamburg hosted the First Hamburg International Environmental Law Conference (HIELC 2011), a law event which, in the editors' opinion, was overdue. The greatest environmental problems are global ones. Key environmental regulations and directives are initiated by the European Commission and adopted by the European Council and the European Parliament, and by means of international agreements, the world community attempts—not always successfully—to address global environmental impacts with global action. Environmental policy, and environmental law as one of its main instruments, is therefore increasingly shaped internationally.

In February 2010, more than fifty legal experts—judiciary, academics, practising lawyers, public servants and policy makers—launched 'Internationaler Umweltrechtstag Hamburg e.V.', a private association established to support the Hamburg International Environmental Law Conference. The association's members together wish to achieve the aim of promoting common understanding at the international level concerning the objectives and capabilities of environmental law and of pushing ahead the development of internationally accepted legal standards in environmental law. The primary means to this end is the exchange of ideas between experts, first and foremost in the framework of the Hamburg International Environmental Law Conferences to be held every two years.

On 15 and 16 September 2011, over 200 lawyers, scientists, students, executive officers and other practitioners from over 30 nations assembled at Bucerius Law School in Hamburg to discuss two of the most urgent environmental problems the globalized world is facing today: Climate change and environmental hazards related to shipping. The event was completed by a town hall reception with Olaf Scholz, First Mayor of Hamburg.

This book contains the presentations held at the First Hamburg International Environmental Law Conference and summarizes the main discussions and results. It begins with the Welcome Address held by Dieter Lenzen, President of Hamburg University, at the town hall reception followed by the opening speech Hans-Joachim Koch gave at the beginning of the first conference day (Part One).

According to the conference's structure, the proceedings are divided into the main topics 'Climate Change' (Part Two) and 'Environmental Hazards

Related to Shipping' (Part Three). The next parts of the proceedings are the descriptions of the additional forums 'Protection of the Marine Environment and Climate Change' (Part Four), Offshore Wind Energy (Part Five) and 'Cities' Contributions to Environmental Protection' (Part Six). The Results summarized by Hans-Joachim Koch (Climate Change) and Doris König (Environmental Hazards Related to Shipping) for the main conference are to be found in Part Seven.

Dirk Messner gave the conference's first presentation ('The Budget Approach—A Framework for a Global Transformation towards a Low Carbon Economy') with an overview of the state of international climate policy. His presentation was followed by Daniel Bodansky's idea of instrumental diversity in climate change law and policy ('A Tale of Two Architectures: The Once and Future U.N. Climate Change Regime'). In her presentation 'Energy Switch in Germany: 100% Renewable Electricity by 2050', Heidi Foth referred to the German energy master plan as an example of an ambitious national action plan. In the afternoon, two working groups deepened the discussion. Working group 1 addressed international climate policy before COP 17 and started off with statements by Marc Pallemaerts ('Legal Options for Regime Evolution in the Climate Change Regime: Some Comments'), Lavanya Rajamani ('International Climate Change Policy: An Indian Perspective') and Jacqueline Peel ('International Climate Law and Policy: An Australian Perspective'). The working group was moderated and summarized by Sebastian Oberthür. The second working group turned to EU and National Initiatives in climate protection. The remarkably different challenges climate protection presents all over the world were demonstrated by the examples of South America (Jorge Caillaux, 'Climate Protection: A South American Perspective') and Japan (Masanori Okada, 'Climate Protection and Sustainability in Japan—Tasks Following March 11th 2011'). This working group was moderated and summarized by Astrid Epiney.

The presentations held on 'Environmental Hazards Related to Shipping' mainly focused on relations and conflicts between international and European regulations. The conference's emissions-related discussions are represented by Tim Bäuerle's presentation ('Integrating Shipping into the EU Emissions Trading Scheme?') and Torben Skaanild's statement from the BIMCO's perspective ('Emissions Reduction and Emissions Trading Systems in Shipping: A BIMCO Perspective'). Bäuerle analyzed the chances and difficulties of integrating the shipping sector into the ETS. Another thematic priority was the prevention of shipping accidents. Lawrence D. Barchue Sr. gave an inside view of IMO's position ('A Cooperative

Compliance Strategy: The Voluntary IMO Member State Audit Scheme'). Alexander Proelss focused on the EU legislation after the *Erika* and *Prestige* oil spills and whether it is in accordance with international law ('The 'Erika III' Package: Progress or Breach of International Law?'). Those questions were further discussed among the speakers and Aldo Chircop ('Some Comments on the Role of Industry in the Prevention of Environmental Hazards Related to Shipping') in a working group. Henrik Ringbom moderated and summarized this session.

The Forum 'Protection of the Marine Environment and Climate Change' drew an impressive picture of the problems related to climate change for marine ecosystems by combining scientific and legal aspects (see Mojib Latif, 'The Impact of Global Warming on the Oceans' and Jochen Flasbarth, 'Use and Protection of the Seas in Times of Climate Change'). Lilly Weidemann gave a short discussion summary for this book.

The forum on offshore wind energy discussed in substantial detail the two principal conflict areas related to generating power offshore: the impacts on the marine environment (Ursula Prall, 'Wind Energy and Marine Environment Protection') and other uses of the sea, especially shipping. Monika Breuch-Moritz presented the key steps that have been taken in Germany to solve those problems in her speech 'Wind Energy and Maritime Spatial Planning'. Ronán Long moderated the forum and gave a detailed summary.

The third forum, 'Cities' Contributions to Environmental Protection', emphasized the local government's role in mitigation and adaptation processes and assembled best practice examples from all over the world. The forum was moderated by Martin Huber and Joachim Sanden, who contributed a theoretical basis (Joachim Sanden, 'Introductory Remarks on Opportunities of the Cities in the Field of Local Climate Change Governance') and the specific results (Martin Huber and Joachim Sanden, 'Conclusions of FORUM II: Cities' Contributions to Environmental Protection').

The proceedings close with the overall conclusions drawn by Hans-Joachim Koch ('Results: Climate Change') and Doris König (Results: Environmental Hazards Related to Shipping).

The views expressed in the articles are those of the authors and should not be taken as reflecting in any way the policies or views of the institutions they are representing.

In place of all those who helped to make HIELC 2011 a success, we would like to thank:

- the City of Hamburg and First Mayor Olaf Scholz for enabling Internationaler Umweltrechtstag Hamburg e. V. to organise the conference;
- our sponsors for all their support (Hamburg Wasser; Rechtsanwälte Klemm & Partner; Latham & Watkins LLP; Nordex SE; Stadtreinigung Hamburg; Universität Hamburg; Vattenfall; Rechtsanwälte Wirtschaftsprüfer Steuerberater Dr. Weiland und Partner; Weissleder & Ewer Notar Rechtsanwälte; Wirtschaftsverband Windkraftwerke e.V., Rechtsanwälte Günther Partnerschaft; Lärmkontor GmbH; E.On Hanse; White & Case LLP; GreenPort; Dalkia GmbH, Carbon & Climate Law Review—Lexxion Verlag; Stiftung Offshore Windenergie, Veolia Umweltservice GmbH; SENATOR Reisen GmbH, Audi AG; Deutsche Bahn AG; Bucerius Law School, Hochtief Solutions AG; Bundesverband WindEnergie e.V.; ETH; Freshfields Bruckhaus Deringer LLP; Graf von Westphalen Rechtsanwälte Partnerschaft; Freie und Hansestadt Hamburg);
- Franziska Bothe and Miriam Jobst for managing the event;
- Fraujansen kommunikation for layout and corporate design;
- Leon Radunovic for booking lots of flights;
- Michél Hennig for setting up www.hielc.org;
- the audience from all over the world for joining the conference;
- Keely Boom for checking most of the non-native contributions in this book;
- the members of Internationaler Umweltrechtstag Hamburg e. V. for their help and ideas; and
- especially Kerstin Gröhn for supporting the conference from preparation until publishing these proceedings and for being the contact person for nearly everyone and every problem.

Great thanks also go to all speakers, panelists and moderators for participating in HIELC 2011 and submitting their contribution to this book more or less on time.

Prof. Dr. Hans-Joachim Koch
Prof. Dr. Doris König
(Scientific Council of HIELC 2011)

Dr. Roda Verheyen and
Prof. Dr. Dr. Joachim Sanden
(Board of Internationaler Umweltrechtstag Hamburg e. V.),

PART ONE

WELCOMING ADDRESSES

WELCOME SPEECH

Dieter Lenzen[1]

Dear First Mayor, dear Minister and Senator, consular corps, dear members of parliament, your excellencies, dear Mr. Kuhbier, ladies and gentlemen:

Hamburg was named European Green Capital in 2011. The award, which the European Union confers to recognize high environmental standards and levels of sustainability, inspired one association to establish an international congress on law. I would like to once again warmly welcome all those participating in the first International Conference on Environmental Law and also to thank the first mayor for inviting us to the City Hall tonight.

The aim of the International Conference on Environmental Law is to provide a forum for legal experts every other year, at which they are able to discuss both the status quo of environmental law and legal guidelines for the future. Environmental matters are rarely a concern for individual states alone and realizing international goals for protecting the environment requires transnational cooperation. This legal conference will make Hamburg a research hub in international environmental law. An additional international network devoted to pursuing environmental questions that arise in the fields of oceanography and climate research is continually growing around the KlimaCampus.

As you may know, Hamburg enjoys a long tradition in the fields of oceanography and climate research. In the late nineteenth century the German Naval Observatory was founded in the city's port. Its main task areas lay in marine meteorology and maritime affairs. Today Europe's largest climate computer is located in Hamburg in the German Climate Computing Center. Universität Hamburg is a co-shareholder. We have been working together with the Max Planck Institute for Meteorology since the mid-1970s and for the past four years have run our own cluster

[1] Prof. Dr. Dieter Lenzen, President of Universität Hamburg. Speech held at the reception for the 1st Hamburg International Environmental Law Conference, on 15 September 2011 at 7:15 pm in the Hamburg City Hall (translated by Gabrielle Robilliard, Universität Hamburg). The spoken word prevails.

of excellence—funded by the German Research Foundation—devoted to climate research.

This has prompted the development over the past few years of the KlimaCampus Hamburg, a place in which important basic research is being carried out. KlimaCampus researchers are working on questions such as: How does the climate system function? How does it react to the CO_2 emissions caused by humans? What prognoses are we able to make? Which adjustments are necessary?

It is not, by the way, just the natural scientists who are working on these questions, but also researchers from the economic and social sciences, media studies, peace studies as well as security experts. This means that we are asking ourselves about the social changes that will result from climate change, the problems—and potentially—the legal issues.

You may have read about it over the last few days in the newspaper: the summer sea ice in the Arctic is diminishing and a shocking new low is threatening to mark the end-of-September records. In addition to this, our researchers have discovered recently a transient large ice-free area the size of Holland in the sea ice. Good for shipping, one could say—in the future we will potentially be able to transport goods in our globalized world along entirely new routes. A north-east or north-west passage skimming the North Pole would not just be quicker and cheaper—it would also save on fuel and thus on the CO_2 emissions so harmful to the climate.

At the same time neighboring states are gazing eagerly at the stores of unrefined natural resources located in these marine areas. Oil and gas are thought to be abundant in this region and would be made accessible by the receding ice. The consequences are conflicts over national territories, over mining rights and rights of use, as well as considerable damage to the environment.

Yet, to whom does the Arctic belong? Which states can and must enact laws to protect this region, which is not only particularly sensitive, but also of central importance for our climate?

One further example of the topics dealt with by the KlimaCampus is the loss of large areas of tropical rainforest and its consequences. Every day around 200 square kilometers of forest—roughly the area between the Altona and Hamburg city halls—disappears. We lose valuable ecosystems in this process. At the same time enormous amounts of CO_2 are released into the atmosphere. In many regions there is legislation in place to regulate deforestation via clear-felling or slashing-and-burning.

The problem lies in policing this legislation in remote and difficult-to-access areas. Those charged with conducting inventories of tropical rainforests thus often experience a nasty surprise when it becomes apparent that underneath the still-intact canopy, most of the trees are gone. As no clear-felling has taken place, satellite pictures are of little use.

Scientists at Universität Hamburg have instead developed a technique allowing them to peer underneath the forest canopy with 'X-ray vision'. And this technique has enjoyed enormous success: the United Nation's Intergovernmental Panel on Climate Change plans to incorporate this "X-ray" method into its body of guidelines so that in the future, countries receive a financial bonus if they are able to demonstrably reduce the destruction of forests.

Yet there is a limit to the efficacy of legal guidelines on environmental and climate protection matters, as a KlimaCampus study on the European Emissions Trading System shows. The idea is sound: the so-called emissions allowances confer an equivalent financial value onto the right to emit the greenhouse gas CO_2. Because the quantity of emissions permitted is small, the allowances are expensive and companies are forced to reduce their emission levels. At least, that is the theory.

Studies conducted by the KlimaCampus, however, show that countries react quite individually to this scheme. While the British view the sale of allowances as an additional source of income, in other nations, unused allowances are being given away for free. In Germany, around three quarters of companies failed to take action during the initial phase of the Emissions Trading System.

Ladies and gentlemen, it is quite obvious that there are a number of concrete points at which the latest findings in scientific research, climate protection and environmental law are closely intertwined. Yet Universität Hamburg is not just active on the research front, but also in teaching and training.

Something that might be of particular interest to you as lawyers: the Max Planck Institute for Comparative and International Private Law and Universität Hamburg have been home to the Max Planck International Research School for Maritime Affairs since 2002. The Research School provides an opportunity for young scientists to work together and participate in an interdisciplinary program focusing on the legal, economic and ecological aspects of climate change—that is, the use, the protection and the legal situation of our oceans.

Ladies and gentlemen, I leave you now to discuss these questions amongst yourselves in greater depth over a glass of wine and a bite to eat. I would like to particularly wish all conference participants—some of whom have traveled great distances to be here—an inspiring exchange of ideas.

OPENING SPEECH

Hans-Joachim Koch[1]

Dear Ladies and Gentlemen,

I would like to welcome you all in the Free and Hanseatic City of Hamburg to the First Hamburg International Environmental Law Conference.

First of all let me thank you for following our invitation, all the more so many of you have made a very long and possibly strenuous journey to Hamburg. We are happy to have you here from all over the world. Considering that—as you are all aware of—global environmental problems can only be successfully addressed by the international community we are happy to welcome participants from the following countries: Japan, The U.S., Canada, Peru, Australia, India, Ireland, The Netherlands, Germany of course, and many others as well.

We are especially happy that we can welcome representatives of important international organisations, which play an important role in the field of global environmental policy, such as the IMO.

With a view to COP 17 in Durban, December this year, the Hamburg International Environmental Law Conference has a good chance of presenting elements of solutions to the main problems in international climate policy: One of the major problems is to find a fair burden sharing solution among the industrial states as well as between industrial states, emergent countries and developing countries.

Moving on to the topic of the reduction of Greenhouse Gases in Shipping, the Conference might be able to address the conflict between the IMO and the EU. Perhaps we can develop a framework on how to integrate the shipping sector in the EU emissions trading scheme.

With respect to the prevention of shipping accidents, the Conference may be able to suggest possible solutions for conflicts between international regulations in the frame of the IMO on the one hand and regional norm setting by the EU and—maybe—even by the states on the other hand. On this point it is worth considering that regional legislation for

[1] Prof. Dr. Hans-Joachim Koch, Hamburg University, Germany.

example by the EU can promote adequate solutions on the international level.

We hope the debates during this conference will be creative, constructive and successful in so far as, at the end, well-founded solutions will be agreed on in the working groups.

I wish you a nice stay in Hamburg and I hope you will find the opportunity to discover our city especially the new quarter being developed in the harbour, the Hamburg Hafen City.

Last but not least the Mayor Olaf Scholz will be delighted to welcome you this evening at 7 o'clock in the city hall.

PART TWO

CLIMATE CHANGE

Presentations

THE BUDGET APPROACH—A FRAMEWORK FOR A GLOBAL TRANSFORMATION TOWARDS A LOW CARBON ECONOMY

Dirk Messner,[1] John Schellnhuber, Stefan Rahmstorf, and Daniel Klingenfeld, J.[2]

Latest research shows that there is only a realistic chance of restricting global warming to 2°C if a limit is set to the total amount of CO_2 emitted globally between now and 2050 (global carbon dioxide budget). We move this global budget to the forefront of our considerations regarding a new global climate treaty in the post-Copenhagen process.[3] Combining findings from climate science and economics with fundamental concepts of equity, the 'budget approach' provides concrete figures for the emissions allowances which are still available to countries, assuming they want to prevent the destabilization of the planet's climate system. Our calculations demonstrate that the time pressure for acting is almost overwhelming—in industrialized countries, but also in emerging economies and many developing nations. We suggest several institutional innovations and rules to manage the global and the national CO_2 budgets in a transparent, fair and flexible way. A sober analysis of the state of the art of climate change science and of the state of multi-lateral attempts to create an effective climate protection accord so far reveals that the budget approach can provide crucial orientation for the negotiations towards a comprehensive post-Copenhagen climate regime. The approach facilitates at the same time an institutional design for a low carbon global economy, setting the necessary incentives for decoupling economic growth from the burning of fossil fuels.

---

[1] German Development Institute.

[2] Reprinted with permission from *Journal of Renewable and Sustainable Energy* 2. Copyright (2010), American Institute of Physics. Dirk Messner, John Schellnhuber, Stefan Rahmstorf, and Daniel Klingenfeld, J. (2010) 2 Renewable Sustainable Energy 2, 14 et seq.

[3] The authors are members of the German Advisory Council on Global Change (WBGU). The WBGU recently published a study on 'Solving the climate dilemma: The budget approach'. This paper builds on the fundamental ideas and findings of the WBGU study and demonstrates that the budget approach could serve as a cornerstone for an institutional design for a global low carbon economy.

1. The 2°C-guardrail Requires Immediate Action

In 1995, the German Advisory Council on Global Change (WBGU) suggested that global warming should be limited to a maximum of 2°C above the pre-industrial level in order to prevent dangerous anthropogenic interference with the climate system.[4] However, in the light of more recent research even a warming of 2°C cannot be regarded as 'safe', but is likely to generate already serious consequences—such as sea-level rise that could render certain island states and densely populated coastal regions uninhabitable. In the 'Copenhagen Accord' from December 2009 the importance of limiting global warming to 2°C or less has been accepted. Many countries have already made the 2°C-guardrail an official goal of their climate policy. Scientists broadly support this climate protection target.[5] A growing number of studies indicate that in a world that has overshot the 2°C-limit, our civilization—which has developed in the amazingly stable climatic conditions of the Holocene—would face unprecedented challenges.[6]

In estimating the level of emissions reductions necessary for compliance with the 2°C-guardrail, various important factors must be considered: these include the emitted quantities of CO_2, but also the emissions of other greenhouse gases, the cooling effect of air pollution (especially sulphur particles), the warming effect of soot particles, the inertia of the climate system (most notably of the oceans), and all the other uncertainties which make a probability analysis the only appropriate assessment. Recent studies show, however, that this complexity can be substantially reduced.[7] Due to the remarkable longevity of CO_2 in the atmosphere, this

---

4 WBGU, 'Scenario for the Derivation of Global CO_2 Reduction Targets and Implementation Strategies', Statement on the Occasion of the First Conference of the Parties to the Framework Convention on Climate Change in Berlin (Special Report 1995).

5 John Schellnhuber et al., *Avoiding Dangerous Climate Change* (Cambridge University Press 2006); Katherine Richardson et al., Synthesis Report Climate Change, Global Risks, Challenges and Decisions (2009), Climate Congress Copenhagen, 10–12. March 2009, Copenhagen: University Copenhagen.

6 WBGU (German Advisory Council on Global Change), 'World in Transition: Climate Change as a Security Risk' (Earths can Publications 2008); Ian Allison et al., 'The Copenhagen Diagnosis. Updating the World on the Latest Climate Science' (2009); World Bank, World Development Report 2010, Development and Climate Change (2009); Oli Brown and Alec Crawford, 'Climate Change and Security in Africa—A Study for the Nordic-African Foreign Ministers Meeting' (2009).

7 Malte Meinshausen et al., 'Greenhouse Gas Emission Targets for limiting Global Warming to 2 °C' (2009) 458 Nature 1158; Myles R. Allen et al. (2009): 'Warming caused by cumulative carbon emissions towards the trillionth tonne' (2009) 458 Nature 1163.

particular substance will become increasingly dominant in the long term in comparison with rather short-lived greenhouse gases and aerosols. For that reason, the trend of atmospheric warming over the course of this century will mainly depend on how much more CO_2 is emitted in total. Specifically, cumulative CO_2 emissions till 2050 will largely determine the extent to which global temperature rise can be kept within the 2°C-guardrail. In order to achieve this with a probability of 67%, fossil CO_2 emissions till 2050 must be capped at around 750 Gt, with only a small residual amount being emitted post-2050. At current emissions rates, however, this CO_2 budget will be exhausted within around 25 years—and even sooner if emissions continue to rise (after a short dip in the wake of the recent economic downturn).

The reversal of the global emissions trend must therefore start as soon as possible; in view of the very limited CO_2 budget, any delay will result in almost unachievable reduction requirements. With a reversal of the trend (and the emissions peak being passed) by 2010, global emissions would need to fall to 50–80% below the 1990 baseline by 2050, with further reductions towards zero emissions being achieved thereafter. Even a slight delay in the reversal of the trend, i.e. postponement of the peak year to 2015, would trigger annual global emissions reduction requirements of up to 5% (as a fraction of 2008 emissions) (Fig. 1). In other words, the world would then have to meet *annual* emissions reduction targets equivalent to those established by the Kyoto Protocol for two full decades! And this would apply not only to the group of Annex-I countries but globally. Delaying the peak year even further to 2020 could necessitate global emissions reduction rates of up to 9% per year—i.e. reductions on an almost inconceivable scale, entailing technological feats and social sacrifices comparable to those of the Allied mobilization during the Second World War. Whatever the details, there is no option but to halt the hitherto unabated rise in CO_2 emissions as quickly as possible and to immediately switch to emissions reductions on a global scale thereafter.[8]

2. The International Climate Policy Context

The current international climate policy context is far from favorable. After the Copenhagen summit and based on the Copenhagen Accord

[8] United Nations Department of Economic and Social Affairs, 'A Global green New Deal for Climate, Energy, and Development' (2009).

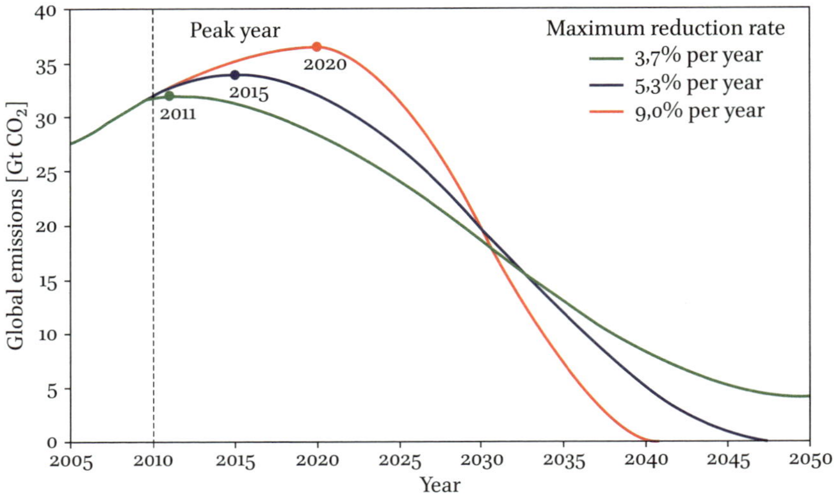

Source: WBGU 2009

Figure 1. Examples of global emission pathways for the period 2010–2050 with global CO_2 emissions capped at 750 Gt during this period. At this level, there is a 67% probability of achieving compliance with the 2°C guard rail. The figure shows variants of a global emissions trend with different peak years: 2011 (green), 2015 (blue) and 2020 (red). In order to achieve compliance with these curves, annual reduction rates of 3.7% (green), 5.3% (blue) or 9.0% (red) would be required in the early 2030s (relative to 2008).

there is a real risk that the UN Climate Change Conference in Mexico at the end of 2010 will only produce a weak compromise which cannot prevent dangerous climate change. According to recent analyses, the various national emissions reduction pledges submitted so far would almost certainly result in warming the planet by significantly more than 2°C.[9] The major actors are moving into the right direction, but far too slowly.

In the climate-negotiations-as-usual process, the world's countries cling to their usual long winded approach, in which complex interests are weighed up in minute detail. The major polluters lay the responsibility at each other's doors: China and other newly-industrializing economies, supported by the Least Developed Countries, point to the high per-capita

[9] Joeri Rogelj et al., 'Halfway to Copenhagen, no way to 2°C' [2009] Nature Reports Climate Change <www.nature.com/climate/2009/0907/full/climate.2009.57.html> accessed 03 July 2009.

emissions produced by the industrialized countries and their emissions-based economic growth since the Industrial Revolution. By way of contrast, the industrialized countries maintain that China is now the largest emitter of GHGs and that emissions in the developing regions, especially Asia, will increase substantially in the future. Finally, the EU stresses that the USA's per-capita emissions are twice as high as its own. The Copenhagen climate conference reflected these beggar-my-neighbor strategies, translating into results at the lowest common denominator. The negotiations appear to have all but stalled now. Therefore a transparent and fair concept for international burden sharing is needed.

3. The Budget Approach—Solving the Climate Dilemma

In support of the international climate negotiations, the German Advisory Council on Global Change developed a new approach that derives national emissions budgets by determining the total ecologically tolerable quantity of global CO_2 emissions up to the year 2050, and apportioning this in line with fundamental principles of equity.[10] Based on a simple, transparent and equitable 'climate formula', all countries' reduction commitments and the requisite financial transfers between industrialized and developing countries are established and calculated on a clear and comprehensible basis.

3.1. *Basic Principles*

We start with the scientific knowledge that in order to keep atmospheric warming below 2°C, the total amount of anthropogenic CO_2 emitted to the atmosphere must be limited.[11] Therefore, the budget approach rests of a binding cap on the total amount of CO_2 that can be emitted from fossil sources up to 2050 (or an alternative meaningful deadline). This ceiling is

[10] See for similar concepts: Sven Bode, 'Equal emissions per capita over time—A Proposal to combining responsibility and equity of rights for post 2012 GHG emission entitlement allocation' (2004) 14 European Environment 300; Development Research Center of the State Council of China, Project Team Climate Change, 'Greenhouse Gas Emissions Reduction. A Theoretical Framework and Global Solutions' in: RossGarnaut et al. (eds), *China's New Place in a World in Crisis* (Brookings Institution Press 2009) 389; J. Pan and Y. Chen, 'Carbon Budget Proposal, Research Centre for Sustainable Development' Chinese Academy of Social Sciences, Beijing (2009).

[11] Malte Meinshausen et al. (n. 7).

an essential prerequisite for ensuring, with a certain level of probability, that the 2°C-guardrail will not be crossed. In this way, humankind would have a specific and defined amount of emissions available. The allocation of this amount would be subject to negotiations; in particular, it can be broken down into national emissions budgets based on equal-per-capita entitlements worldwide.

Proposals for the allocation of emission rights have particularly good prospects of being accepted by the international community if they are viewed as fundamentally equitable by as many stakeholders and affected parties as possible. In this context, we propose that the allocation should be based on three 'first principles', namely the polluter-pays principle, the precautionary principle, and the principle of equality.

In accordance with the *polluter-pays principle*, the industrialized countries have a particular responsibility to cut their greenhouse gas emissions due to their high cumulative emissions in the past. Unless the industrialized countries act according to this responsibility, no global climate treaty will hardly ever be achieved.

In line with the principle of sustainability[12] and based on the 2°C-guardrail, the *precautionary principle*[13] must be respected; this means that timely action is required to prevent irreversible damage to present and future generations. The global emissions budget as induced by the 2°C guardrail requires not only the industrialized countries but also the emerging economies and developing countries to adopt a course towards a low-carbon future. 'Catch-up' development in Africa, Asia and Latin America during the 21st century that is based primarily on fossil fuels would gamble with many of humankind's natural life-support systems.

Conversely, the populations of the industrialized countries do not have a natural right to per-capita emissions many times greater than those of the developing countries. The *principle of equality*—which postulates individuals' equal rights, without distinction, to the benefits of the global commons—is recognized by many countries but is not yet enshrined in law. The UN General Assembly[14] and the UNFCCC[15] acknowledge 'that change in the Earth's climate and its adverse effects are a common concern of humankind'. From a theory-of-justice perspective, this concern does

[12] Article 3, Paragraph 4, first sentence UNFCCC.
[13] Article 3, Paragraph 3 UNFCCC.
[14] Resolution 43/53, 1989.
[15] See Preamble of the UNFCCC.

not permit any differentiation based on national or individual interests.[16] It requires emissions to be allocated in a manner which reflects the interests of the global community and humankind as a whole. The principle of equality does imply that equity in per-capita emissions should be the basis for the allocation of national emissions budgets.

Due to the socio-economic conditions, the global budget allocated on a country-by-country level cannot be utilized entirely without time constraints, as the 2°C-line can only be held if realistic decarbonization dynamics are taken into full account. Firstly, it takes decades to restructure emission-intensive infrastructures and production processes and to transform consumer behaviour, so there is no easy way to reduce global emissions at high speed. Secondly, global decarbonization must commence as soon as possible, as the best of our current knowledge indicates that any postponement now would necessitate emissions reductions at an unattainable rate in future. Thirdly, at the end of the budget period, i.e. around 2050, a virtually zero-emissions economy must have emerged, as the remaining environmental space for the subsequent decades is likely to be very limited: the accumulation of CO_2 in the Earth System by the middle of this century would be just about tolerable, yet the high carbon load would persist for a long time due to the geophysical and biochemical processes and peculiarities involved.

In order to accommodate these constraints, the budget approach must therefore be fleshed out with specific rules:

1. *Global interim targets*: As an important milestone, it should be stipulated that the global CO_2 emissions curve must peak between 2015 and 2020 and decline thereafter (see Fig. 1). As a further milestone it should be stipulated that by 2050 full decarbonization is to be widely accomplished.
2. *National decarbonization road maps and interim targets*: All countries should pledge to develop and present national strategies to manage their own CO_2 budgets ('decarbonization road maps'). These should be based on realistic evaluations of the national emissions reduction potentials as a function of time; their plausibility and operability should be verified by an independent international institution. This would reduce the risk that certain governments would postpone the necessary action indefinitely and leave all burdens to future generations. By

[16] John Rawls, *A Theory of Justice* (Cambridge University Press 1971).

allowing high flexibility in the choice of transformation pathways, the strengthening of countries' individual responsibility is coupled with accountability to the international community.

3. *Interregional flexibility*: The unrestricted and efficient management of national budgets via a global CO_2-emissions trading system is highly recommended. As a prerequisite, the national budgets must be declared to be tradable quota. International emissions trading allows and encourages a very wide range of bilateral and multilateral transactions. For example, trading permits industrialized countries, who have used up almost all of their CO_2 budget, to purchase allowances, but also enhances the incentives for countries to reduce their own emissions. Substantial capital flows are generated for the developing countries and here too, incentives are created for emissions reductions, as CO_2 budget surpluses can be traded and monetized.

3.2. *Calculating National Emission Budgets*

For the proposed budget approach, the total available global emissions budget of CO_2 from fossil sources that allows compliance with the 2°C-guardrail is calculated for a specific period. This amount is then allocated among the individual countries, based on equal cumulative per-capita emissions over a fixed period.

It is then very easy to determine the national budgets because the model only contains *four*—political, i.e. negotiable—parameters: the period for the total budgeting as defined by a *start* year and an *end* year; the *probability* of achieving compliance with the 2°C-guardrail, and finally, a *demographic reference year* for the national population strengths that form the basis for the concrete allocation figures. We propose that the year 2050 be selected as the end point for the budget period, as there is no doubt that most of the requisite emissions reductions will need to have been achieved by that date.

The other three parameters determine the overall budget amount and its relative distribution. They are therefore parameters of the utmost political relevance.

- The *start year* determines the point at which global equitable emissions management—in other words, the proposed distribution formula—should come into effect. If the start year is backdated (i.e., a year in the past is defined as the commencement point), the budget to be distributed will include emissions that have already taken place. This inevitably means that countries with high historical per-capita emissions will have a proportionately smaller emissions budget available in the future.

- Due to the complexity of the climate system, it is not possible to calculate *precisely* what additional amount of CO_2 can still be absorbed by the Earth's atmosphere without breaching the 2°C-guardrail. However, with the help of so-called ensemble calculations and using the best simulation models available worldwide as well as sophisticated statistical methodologies, climate science can compute the *probabilities* that a specific total amount of emissions will remain 'sub-critical' and will thus be compatible with restricting warming to 2°C. The global budget that is available for distribution can thus be calculated directly on the basis of a political global risk assessment: the higher the probability of compliance with the 2°C-guardrail that is chosen, the smaller the budget to be spent world-wide over the chosen period must be. It is crucial to note at this point that probabilities of averting damage that fall within the 50–90%-range—i.e., the range generally discussed in relation to the climate problem—would be considered completely unacceptable in every-day contexts (e.g. with respect to traffic safety, prevention of infectious diseases, etc.)! Unfortunately, global environmental change has progressed too far already for a genuine precautionary policy that satisfies the criteria of common sense.
- The *demographic reference year* determines the national share of the global budget based on the country's relative demographic weight for the given year. The later the reference year that is chosen, the higher the benefit for countries with rapidly growing populations as their relative demographic weight is steadily increasing and, concomitantly, their relative cumulative-emissions share. Instead of accounting for the population figures in a single year, allocation could also be based on the (mean) population weight over a longer period, involving projections of trends. The argument in favour of using a fixed reference year is that in countries with high population growth rates, it could serve as an incentive for demographic change, since unabated population growth after the reference year would stretch the allocated budget, effectively reducing emissions per capita. This reflects the fact that the absolute amount of cumulative emissions world-wide is limited, so per-capita allowances shrink with a growing global population.

In a broader perspective, the outlined budget approach serves to analyze the low-carbon transformation challenge from four angles:

- Taking stock with respect to national emissions budgets—defining the magnitude of the transformation challenge;
- Developing the foundations for a new global climate architecture—simplifying the negotiation process and focusing on environmental integrity;
- Proposing an institutional design for a low-carbon world economy—setting incentives for a deep transformation process;
- Outlining new climate and development partnership patterns among high-emitting and low-emitting countries.

3.3. *Taking Stock Based on National Emissions Budgets*

The budget approach allows to discuss—in quantitative terms—different policy options based on different decisions regarding the four parameters of the 'climate formula' (see Table 1). Here we outline one of the evident (although by no means self-evident) options, termed 'Future Responsibility'. It already constitutes a compromise between what is scientifically necessary, on the one hand, and politically and economically feasible, on the other: the probability of holding the 2°C-line is set at 67% and the global emissions budget is apportioned based on the current reference year (i.e. 2010). The historical responsibility of developed countries for past emissions would not be considered explicitly in the national budgets available. Therefore, additional compensation mechanisms in developing countries, particularly support and financial resources for adaptation measures, could help bring about a more politically acceptable, yet reasonably equitable agreement.[17]

Table 1 shows examples of the national emissions budgets, based on a permissible global budget of 750 Gt CO_2 from fossil sources between 2010 and 2050 that is allocated to the countries on an equal per-capita basis. In this option, the start year 2010 would appear to be a suitable demographic reference year. As a consequence, every person (based on the 2010 world population) would be allocated a budget of around 110 t CO_2 emissions for the next 40 years, equivalent to average annual per-capita emissions of approx. 2.7 t CO_2. However, in order to safeguard the transition to those low emissions that are permissible post-2050, global mean per-capita

[17] WBGU, 'Solving the Climate Dilemma. The Budget Approach' (Special Report 2009) 33. We also calculated a "Historical Responsibility" scenario, with a 75% probability of compliance with the 2°C guard rail, adopting 1990 at the start year for the budget period, ibid., 25–26.

Table 1. Allocation option 'Future responsibility', 2010–2050; 67% probability of compliance with the 2°C guard rail; 2010 as the reference year for population data. Only includes CO_2 emissions from fossil sources. CO_2 emissions for 2008 and population numbers for 2010 are estimations.

	Share of world population in 2010 [%]	Budget 2010–2050 [Gt CO_2]		Estimated emissions in 2008 [Gt CO_2]	Reach of the budget lifetime, assuming annual emissions as in 2008 [years]
		Total period	Per year		
Germany	1.2	9.0	0.22	0.91	10
USA	4.6	35	0.85	6.1	6
China	20	148	3.6	6.2	24
Brazil	2.8	21	0.52	0.46	46
Burkina Faso	0.24	1.8	0.043	0.00062	2892
Japan	1.8	14	0.34	1.3	11
Russia	2.0	15	0.37	1.6	9
Mexico	1.6	12	0.29	0.46	26
Indonesia	3.4	25	0.62	0.38	67
India	18	133	3.2	1.5	88
Maldives	0.0058	0.043	0.0011	0.00071	61
EU	7.2	54	1.3	4.5	12
World	100	750	18	30	25

Source: WBGU 2009, using data from: Meinshausen et al., 2009; WRI-CAIT, 2009; U.S. Census Bureau, 2009

emissions at the end of the budget period must lie well below this average: by 2050, global CO_2 emissions must be around two thirds lower than the 1990 baseline despite a growing world population. This would result in an entitlement to annual per-capita emissions of around 1 t CO_2 by 2050! Fig. 2 shows selected countries and the mean annual budget to which they would be entitled, as compared with their respective emissions in 2005.

The results depict a dramatic state of affairs for the USA, Germany, and the EU: for instance, at current emissions levels the German CO_2 budget allocated according to the option discussed would be exhausted within 10 years. In the light of this situation, it is clear that the industrialized countries must carry out a rapid and comprehensive decarbonization of their economies by 2050. However, even drastic domestic reduction efforts will not be enough to keep OECD countries within their budget constraints. It is remarkable that within this scenario, the situation for China (and similar for Mexico, Argentina, Chile or Thailand) is also rather challenging. If China maintains its emissions on a 2008 level, its budget would only last for some 24 years—a similar observation applies for the world as a whole.[18]

On the other hand, it is neither possible nor necessary for all countries to tackle mitigation solely through their own domestic emissions reductions efforts. The allocation scheme proposed is equitable, yet should be regarded as the basis for effective and efficient burden-sharing. This can take place in a variety of ways: alongside domestic emissions reductions, the trading of emission allowances is likely to play a key role, but other flexible mechanisms for international climate cooperation, along with financial and technology transfers, can help to speed up progress towards low-carbon development all over the world.

3.4. *Foundations for a New Global Climate Architecture*

The budget approach has the potential to make two critical contributions to the construction of a global climate architecture, namely to maintain the environmental integrity of a new climate regime while breaking the negotiation stalemate between the major global emitters. This 'Gordian Knot of climate policy'—a knot tied mainly by the USA, the EU, China and the G77 countries—can be described in terms of game theory as a 'social dilemma': rationally justified individual preferences lead collectively to an outcome whereby all players end up worse off in the future, and furthermore—in the case of climate change—sustain massive and irreversible damage to the Earth's life-support systems. Unless the key players rise above the tactics of self-interest at the ongoing and forthcoming climate negotiations, 'dangerous anthropogenic interference with the climate system' (Article 2 UNFCCC) can no longer be prevented.

A further factor impeding the negotiations is that even an agreement on ambitious and binding greenhouse-gas reductions between the industrialized countries and the populous or fast-growing newly industrializing economies is no longer enough to keep warming below the 2°C-limit. Many developing countries are now achieving high dynamic economic growth, accompanied by rising fossil fuel consumption. For that reason,

---

[18] All figures refer to fossil CO_2 emissions (i.e. without land-use changes). The emissions from land-use changes between 2010–2050 are estimated at around 60 Gt CO_2. We consider these additional emissions in our calculations of future climatic warming. But in relation to land-use changes, we suggest that a separate regime is likely to be more successful. The key issue is that allocating the national emissions budgets creates a shared responsibility for climate change mitigation.

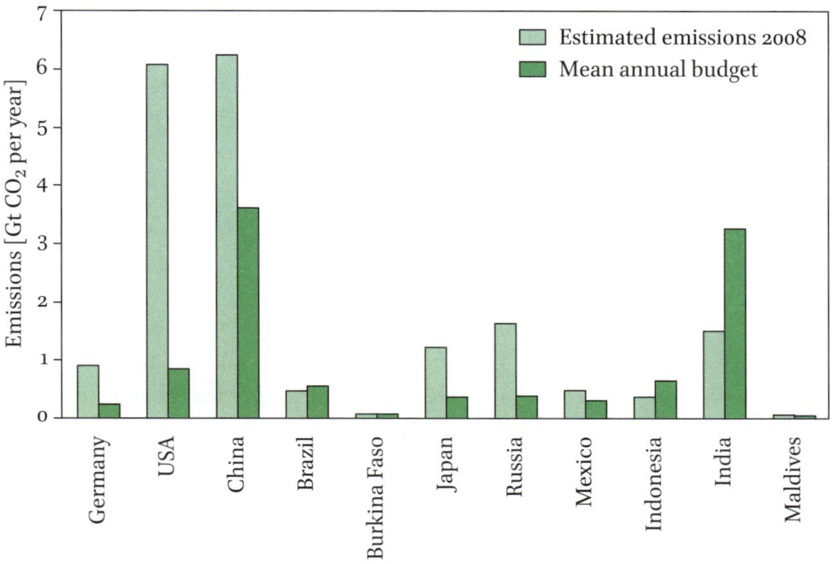

Source: WBGU 2009

Figure 2. Fossil CO2 emissions in 2008 (estimated figures) and permissible average annual budgets under allocation option 'Future responsibility' for selected countries.

a course must be set on a global basis towards a low-carbon economy. It is clear that even for the majority of developing countries, doggedly persuing fossil-fuel-driven growth is no longer an option if dangerous climate change is to be averted. Economic and social development must be decoupled from greenhouse gas emissions worldwide.

But any future attempt to accomplish this based on the business-as-usual model of international climate policy would require all 194 signatories (as of December 2009) to the UNFCCC to negotiate emissions stabilization and reduction targets for around 100 countries! This is the group of countries that already is emitting more than the yearly emissions allowance of around 2.7 t CO_2 per capita for the 2010-to-2050 budget period. The process is at risk of deteriorating into a 'talking shop' in which negotiators haggle over every decimal point and comma at marathon bargaining sessions even as global warming spirals out of control. Against this background, the budget approach presents the basic structure for a global architecture for climate protection, and shows how compliance with the 2°C-guardrail can be achieved. The proposal that countries agree on a simple and equitable 'climate formula' as the basis on which all countries can be allocated a precisely determined national emissions budget in the future is the fundamental concept behind the global climate compromise. As a result, the budget approach reduces the complexity of the climate negotiations and thus identifies a way forward for the climate negotiation process which helps countries to move on from their narrow national agendas.

Based on this analysis, we conclude that parties to the UNFCCC will have to agree on the following general principles in the next negotiation rounds:

- The *2°C-guardrail* is adopted as legally binding in international law.
- For carbon dioxide—the greenhouse gas crucial to climate protection efforts in the long term—a *global emissions budget* for fossil sources up to the year 2050 that is compatible with the 2°C-guardrail is adopted on a legally binding basis as well.
- The following *milestones* are stipulated: (1) The peak year of worldwide CO_2 emissions is to be reached between 2015 and 2020; (2) global emissions by mid-century are to be reduced to a level consistent with the narrow emissions budget remaining post-2050.
- The global CO_2 budget is distributed among the world's population on an equal per-capita basis so that *national* CO_2 *budgets* can be calculated for all countries, and adopted on a legally binding basis. These budgets provide orientation for countries on how swiftly and substantially their CO_2 emissions need to be reduced.
- Each country is committed to producing internationally and objectively verifiable *decarbonisation road maps,* which provide information on the planned national emissions path up to the year 2050. These road maps should be based on the national CO_2 budgets as well as on the respective national emissions reduction potentials.
- In addition, for the countries with presently high per-capita emissions, *reduction commitments up to 2020* are agreed in order to avoid delaying decarbonisation efforts.
- *Flexible mechanisms* (international emissions trading and Joint Implementation) as well as appropriate additional financial and technological transfers by the industrialized countries are agreed upon.
- The *separate regulation* of CO_2 *from non-fossil sources, other relevant greenhouse gases and future radiative-forcing substances* is to be pursued intensively as it creates opportunities for swift reductions in total emissions harmful to the climate.

4. The Institutional Design for a Low-Carbon Global Economy

The budget approach does not only serve as a compass for the future climate regime, it provides at the same time the fundamental elements of an institutional design for a global low-carbon economy. The approach provides great transparency as it explicitly reveals the global and national CO_2 scarcities that have to be accounted for in the creation of public and private welfare. The pertinent scarcities are actually of a character that should compel many countries to develop low-carbon strategies without further delay. Moreover, the obvious quantitative limitations as reflected in the global and the national budgets, respectively, create clear incentives for the transition towards a trans-fossil era. The budgets define the space in which the economic actors need to accommodate their strategies and activities. For the private sector, this transparency and the long-term constraints agreed upon result in stability, anticipatory certainty for the whole budget period from 2010–2050, a global level playing field, and a clear system of incentives for investments in the future and low-carbon innovation efforts. Within the framework of the budget approach, every success in reducing emissions is rewarded equally, regardless of the country in which it is achieved. CO_2-emissions reductions translate into a form of capital. Climate protection and carbon efficiency become key factors in boosting competitiveness. International competition for the most innovative decarbonization strategy, i.e. a race to the top, can finally start to roll.

We have demonstrated that the high-emitting countries will not be able to meet their climate challenges within their own national budgets. When implementing the budget approach, the industrialized countries in particular would need to decide to what extent they wish to undertake the necessary greenhouse gas reductions at home or would rather make use of flexible mechanisms (emissions trading, Joint Implementation) which permit cost-efficient fulfilment of national reduction commitments. Emissions trading will therefore be a key institutional element of a global low-carbon economy and a major driver of low-carbon innovation.

How would the trading patterns between countries look like? Three major groups can be distinguished (Fig. 3) according to the number of years which their allocated budgets can be expected to last (excluding emissions trading):

- Group 1: Countries whose budget—at their current rate of emissions—would be exhausted in *less than 20 years*. Under the assumption of a linear reduction trajectory, these economies (around 60 countries including the EU, US, Saudi Arabia, Venezuela, and Malaysia) would therefore have to achieve zero emissions before 2025/30 if they are to stay within the budget.
- Group 2: Countries whose budget—at their current rate of emissions—would last for at most *20–40 years* (around 30 countries including China, Mexico, Chile, and Thailand), i.e., roughly till the end of the global budget period in 2050.
- Group 3: Countries whose budget—at their current rate of emissions—would last for *more than 40 years* (some 95 countries in total including the least developed countries in Subsaharan Africa, Vietnam, and India), i.e., beyond the end of the global budget period.

For all our calculations, assumptions for the 'current rate of emissions' are based on estimated values for 2008. Group 3-countries would be the main suppliers of allowances, as even Group 2-countries would probably only be able to generate small emissions surpluses over the entire budget period. Emissions trading would create novel prospects for income generation for the suppliers of emission allowances. For climate policy to succeed, it would be essential that those Group 3-countries that do not fully exploit their emissions budget (even when experiencing rapid economic growth) invest their revenues from the sale of allowances in low-emissions technologies rather than remaining on 'fossil trajectories'. Thus not only national decarbonization milestones should be set; there should also be at least a partial earmarking of revenues from inter-country emissions trading. In such a scheme, a substantial proportion of the so-generated revenues should be invested in low-emissions technologies, especially into energy production from renewables sources.

Joint Implementation (JI) is a further flexible mechanism, established by the Kyoto Protocol, and also relevant for the budget approach. JI allows countries subject to reduction commitments to create additional domestic emission rights by carrying out emissions reduction projects in other countries that have also accepted commitments to limit emissions. Within the budget approach, Joint Implementation would create incentives for industrialized and other high-emission countries to invest in mitigation technologies in developing countries and emerging economies, i.e., countries belonging to Group 2 or Group 3. For countries in those groups that

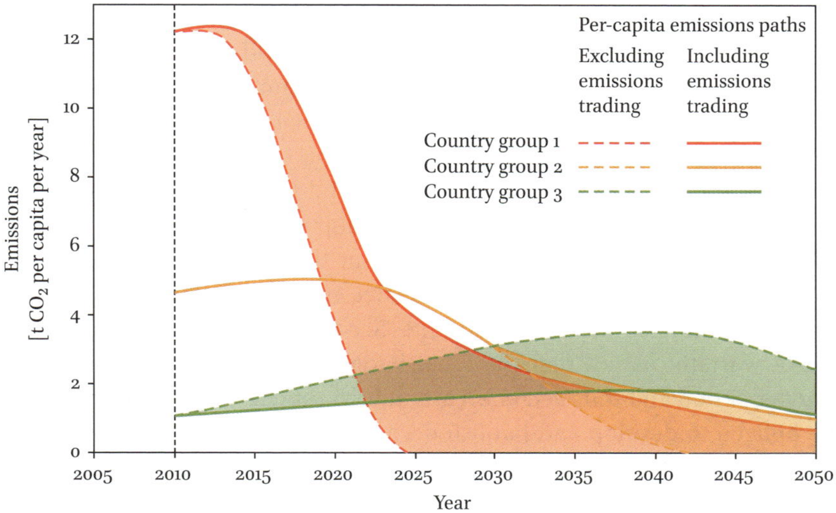

Source: WBGU 2009

Figure 3. Examples of per-capita emissions paths of CO_2 from fossil sources for three groups of countries according to the WBGU budget approach, which could emerge through emissions trading (*unbroken curves*). Here it is assumed that the countries of group 1 (> 5,4 t CO_2 per capita per year) will raise their budgets by 75% by purchasing emission rights for 122 Gt CO2. The countries in group 2 (2,7–5,4 t CO_2 per capita per year) purchase additional emission rights for a total amount of 41 Gt CO_2. The countries of group 3 (< 2,7 t CO_2 per capita per year) become sellers of a total of 163 Gt CO_2 and, accordingly, their budget sinks by approximately 43%. Towards the end of the budget time period there is a convergence of the actual CO_2 emissions at approximately 1 t CO_2 per capita per year (relating to the population in 2010). The *broken curves* show the theoretical per-capita emissions paths for CO_2 without emissions trading. The areas between the curves illustrate the traded amount of emission allowances. Due to the fact that the illustration shows the per-capita situation and the country groups have varying sizes of population, the areas between the purchasing country groups 1 and 2 do not coincide with the area of the selling country group 3.

are unable to sell emission rights, JI is a promising mechanism by which to finance their transformation processes. The basis for this very mechanism would be substantially broadened in comparison to the Kyoto Protocol, because national CO_2-emissions budgets would be allocated to all states, with the Clean Development Mechanism (CDM) being subsumed under Joint Implementation.

To make the budget approach and the global emissions trading scheme work, we recommend the establishment of a World Climate Bank.[19] The World Climate Bank would have several important tasks. Firstly, it would be responsible for the monitoring of the actual emissions of countries and the imposition of sanctions. To ensure, in a systematic manner, that budgets are not exceeded, it would be helpful to set national milestones for manageable time periods and to develop other corresponding rules. The sanction rules should be publicized at the beginning of the budget period and should be sufficiently effective to create strong incentives to remain within the budget constraints. Secondly, to avoid that countries make wasteful use of their national budgets in the early phases of the budget period and thus become 'carbon bankrupt' later, all nations should be obliged to develop and publish decarbonization roadmaps. They have to be assessed by and improved with the help of the World Climate Bank. Thirdly, The World Climate Bank should play an important role in shaping Joint Implementation partnerships. Fourthly, the bank would need to monitor the development over time of the national (and thus of the global) emissions, in order to ensure compliance with the 2°C-guardrail. In particular, it is important to make sure that global CO_2-emissions start to decrease at some point during the period from 2015 to 2020 and reach very low levels towards the year 2050. In order to accomplish that global CO_2 emissions actually peak prior to 2020, it would be expedient to structure the use of the flexible mechanisms by introducing trading periods. The World Climate Bank, as an institutional innovation, would form a 'climate pillar' of the revised global economic governance architecture, becoming as important as the Bretton Woods institutions (International Monetary Fond, World Bank) and the World Trade Organization.

5. New Climate Partnership Patterns

If the world succeeds in setting a course towards a global low-carbon economy via the budget approach, then the national actors will need to stick closely to the following script: More than 100 countries, whose emissions are already today beyond the 2.7 t-CO_2-per-capita-emissions amount that

[19] For a similar concept see Lutz Wicke, *Beyond Kyoto—A New Global Climate Certificate System. Continuing Kyoto Commitments or a Global 'Cap and Trade' Scheme for a Sustainable Climate Policy*? (Springer 2005).

would be available on average between 2010 and 2050, should initiate, as quickly as possible, a transformation process which allows them to stabilize their emissions before long and then to move towards almost complete decarbonization. Only 65 countries have economies whose emissions are currently below 1–1.5 t CO_2 per capita per year and are thus (mostly nolens-volens) climate friendly. The CO_2 budgets of the industrialized countries are extremely limited. These countries should therefore combine strategies for the radical restructuring of their 'fossil' economies with the use of flexible mechanisms (such as technology transfer to reduce greenhouse gas emissions in developing countries) and the purchase of substantial quantities of emission allowances. Many emerging economies also need to make substantial decarbonization efforts if they are to stay within their budgets by mid-century without having to purchase additional emission certificates. The majority of emerging economies are therefore highly unlikely to become suppliers of tradable allowances. This gives developing countries with consistently low levels of emissions an overwhelming strategic importance, as owners—and sellers—of such emission rights. The budget approach offers them an unprecedented opportunity to accelerate their future development through technology and financial transfers, and to move towards a low-carbon industrial metabolism from the outset. It becomes clear that the budget approach binds together climate policies, the creation of essential elements of on institutional design for a climate-efficient economy, and mechanisms for a global low-carbon development partnership.

From the above clusters of interests, it becomes apparent, *firstly*, that an historic climate partnership between Group 1-countries (essentially the industrialized countries) and Group 3-countries (essentially today's poorest countries) is vital in solving the climate problem. They will operate on the principle of technology and financial transfers in exchange for 'budget surpluses'. The 'donors' and 'recipients' who have traditionally been the key actors in development cooperation thus become partners with mutual common interests.[20] International climate change mitigation must thus go hand in hand with a global development partnership between

[20] Dirk Messner, 'Four Lessons from the Present Global Financial Crisis for the 21st Century', in Ashwani Kumarand Dirk Messner (eds), *Power Shifts and Global Governance. Challenges from North and South* (Anthem Press 2009), 42.

'high-emission countries' and 'low-emission countries'.[21] For many developing nations, these transfers hold the key to a zero-emissions future, as the development of their sustainable energy systems could largely be funded through emissions trading. The dual benefit is that these countries can avoid the burden of fossil path dependency without forfeiting their opportunities for development.

A more detailed analysis of the potential suppliers of emission allowances generates the following, fairly realistic scenario: sub-Saharan Africa is the region with the largest number of countries that could sell their surpluses. However, many of the suppliers in Group 3 are small economies which—from a potential purchaser's perspective—have only very modest amounts of greenhouse gas emissions available to sell. From the perspective of the Group 1-countries, the more attractive potential suppliers with large volumes of emission allowances are India (population: 1.2 billion; budget would last for 88 years if emissions remain constant, see Table 1), Bangladesh (population: 164 million; 384 years), Pakistan (population: 185 million; 125 years) and Ethiopia (population: 85 million; 1251 years). Although their geostrategic role is negligible at present, climate change issues in the future could make them major players with whom the industrialized countries should maintain particularly constructive relations.

Secondly, it is clear that emissions trading between Group 1-countries and Group 2-countries is likely to be very limited due to the minimal or modest budgets available to both groups. Nonetheless, the industrialized countries are likely to have great interest in technology partnerships on equal terms in order to reduce the greenhouse gas intensity of the newly-industrializing countries (especially China) and thus to prevent these countries from becoming major purchasers of emission allowances themselves. Competition between Group 1 and Group 2 for Group 3's limited supply of emission allowances would undoubtedly drive up prices. This insight could encourage the formation of climate alliances between China, the EU and the USA, for example. Despite such common interests, however, there is likely to be intense future competition between the industrialized countries and China in particular regarding technological leadership during the phase of global decarbonization.

---

21 Ottmar Edenhofer et al., *The Economics of Decarbonization* (Potsdam Institute for Climate Impact Research 2009), 43 et seq.; Leggewie, C./D. Messner (2012): The low carbon transformation—a social science perspective, Journal for Renewable and Sustainable Energy, No. 4, 2012, doi: 10.1063/1.4730138 (added in October 2012).

Thirdly, our analysis shows that the two most important emerging economies in Asia, namely China and India, will be confronted with very different challenges under the budget approach. China, due to its impressive economic growth dynamics and an already highly level of per-capita emissions, will need to implement a comprehensive decarbonization strategy with great urgency (McKinsey, 2009).[22] India, by way of contrast, due to its still relatively low per-capita emissions, has the chance to pursue a gentler transformation pathway towards a low-carbon economy, despite its high economic growth potential: In other words, India will face far less time pressure than China in starting to decouple economic development from greenhouse gas emissions. The sooner India recognizes this opportunity to avoid 'catch-up' carbonization, the greater its prospects of becoming a major supplier of emission allowances in future. India could thus greatly facilitate its pathway towards a low-carbon future through partnership with the industrialized countries. Equally, climate cooperation with India would be of great strategic interest for the North, especially for securing its own access to India's emission allowances. In summary, India could become a key actor and major winner if a world climate treaty would be adopted in line with the budget approach after all.

[22] McKinsey, *Pathways to a Low-Carbon Economy. Version 2 of the Global Greenhouse Gas Abatement cost Curve* (McKinsey & Company 2009).

A TALE OF TWO ARCHITECTURES: THE ONCE AND FUTURE U.N. CLIMATE CHANGE REGIME

Daniel Bodansky[1, 2]

Ever since the Kyoto Protocol's entry into force in 2005, the central question facing the United Nations climate change regime has been what to do after 2012, when the Kyoto Protocol's initial limits on greenhouse gas emissions expire. When states negotiated the Kyoto Protocol more than a decade ago, their intent was to establish an enduring policy architecture, consisting of internationally defined, legally binding emissions reduction targets, combined with market mechanisms such as emissions trading to achieve those targets. Although the Kyoto Protocol itself set targets for only a five-year commitment period running from 2008–2012, the expectation was that this first commitment period would be followed by a second commitment period, a third, and so on, indefinitely into the future.

But relatively few countries, representing only about a quarter of global carbon dioxide emissions, have been willing to accept internationally defined emissions reduction targets under the Kyoto Protocol.[3] And even

---

[1] Lincoln Professor of Law, Ethics and Sustainability, Sandra Day O'Connor College of Law Arizona State University A. B., Harvard, 1979; M. Philin the History and Philosophy of Science, Cambridge, 1981; J.D., Yale 1984. Professor Bodansky served as the Climate Change Coordinator in the U. S. Department of State from 1999–2001. This article draws on the author's previous work, including 'Targets and Timetables: Good Policy but Bad Politics?', in Joseph E. Aldy and Robert N. Stavins (eds), *Architectures for Agreement: Addressing Global Climate Change in the Post Kyoto World* (Cambridge University Press 2007); 'The Future of Climate Governance: Creating a More Flexible Architecture', in Richard Stewart et al. (eds), *Climate Finance: Regulatory and Funding Strategies for Climate Change and Global Development* (New York University Press 2009) 48; 'The Copenhagen Climate Change Conference: A Postmortem', (2010) 104 Am. J. Int'l L. 230; 'The International Climate Change Regime: The Road from Copenhagen', Viewpoints, Belfer Ctr. For Sci. and Int'l Affairs, Harvard Project on Int'l Climate Agreements, Kennedy Sch. of Gov't, Harvard Univ., Cambridge, Mass. Oct. 2010 <http://belfercenter.ksg.harvard.edu/files/Bodansky-VP-October-2010-3.pdf> accessed 05 July 2012; and with Elliot Diringer, 'The Evolution of International Regimes: Implications for Climate Change' (2010) <http://www.pewclimate.org/docUploads/evolution-multilateral-regimes-implications-climate-change.pdf> accessed 05 July 2012. The author is thankful for very helpful comments from Elliot Diringer and Susan Biniaz.

[2] Reprinted with permission from Arizona State Law Journal, volume 43, issue 3, pp. 697–712 (2011).

[3] Although the Kyoto Protocol now has 193 parties, it establishes emissions targets for only 38 countries (plus the United States, which has not ratified the Protocol).

some of these, such as Japan, Russia, and Canada, seem unwilling to continue down the Kyoto path. The other main emitters—most notably the United States, China, and India—have never accepted that they should be subject to emissions targets established from the top-down, through international negotiations. Instead, they have preferred to address the climate change problem, if at all, from the bottom-up, through nationally determined policies such as efficiency standards and technology initiatives.

Going forward, the climate change regime faces a choice: continue down the road blazed by Kyoto, or shift to a more bottom-up architecture, focusing on nationally defined measures. Countries differ widely about these two policy architectures. The European Union would prefer to stick with the Kyoto architecture, but broadened so as to encompass the emissions of the other major economies—in particular, the United States and China. The big developing countries would like developed countries to accept a second round of emissions targets under the Kyoto Protocol but are unwilling to do so themselves. Meanwhile, the United States does not appear to care one way or the other whether the countries with Kyoto targets negotiate a second commitment period for themselves, and has instead focused on advancing a more bottom-up architecture to which it could subscribe.

To date, the U.N. climate change negotiations have failed to resolve this fundamental debate about the regime's long-term architecture. The 2009 Copenhagen Accord sets forth a bottom-up architecture based on national pledges, and the 2010 Cancun Agreements brought this approach into the official U.N. Framework Convention on Climate Change (UNFCCC) process. But both agreements are explicitly without prejudice to the long-term architecture of the regime, and leave open the possibility of a second commitment period under the Kyoto Protocol (although the prospects for negotiating such an extension appear slim). The battle over policy architectures will therefore continue at this year's climate conference in Durban, South Africa.

This paper explores the development of the climate change regime, the recent meetings in Copenhagen and Cancun, and the options for a long-term architecture going forward.

1. The Origins of the U.N. Climate Change Regime

Although international agreements generally involve negotiations—and all ultimately depend on state consent—they vary widely in the latitude that they give participating countries. Some take a top-down approach, defining particular policies and measures that parties must undertake. Others adopt a more bottom-up approach, allowing each participating state to define its own commitments. In the environmental realm, the Convention on International Trade in Endangered Species (CITES)[4] illustrates the top-down approach. It prescribes which species to protect and how to do so (through a permitting system for imports and exports).[5] Similarly, the international oil pollution treaty (MARPOL) prescribes very specific rules regarding the construction, design, and performance of oil tankers.[6] Conversely, the U.S.-Canada Air Quality Agreement illustrates a more bottom-up approach, largely codifying in an international agreement the domestic air pollution programs of the two participating states.[7] Similarly, the Ramsar Convention on Wetlands encourages countries to promote the conservation and 'wise use' of wetlands,[8] but gives countries broad discretion to determine the policies and measures that they will use to do so.[9]

In the development of the international climate change regime, the top-down approach adopted by the Montreal Protocol on Substances that Deplete the Ozone Layer[10] initially served as a major inspiration for many countries. The basic regulatory approach of the Montreal Protocol is to impose quantitative limits on national production and consumption of ozone-depleting substances (ODS). Originally, the Montreal Protocol required parties to freeze and then reduce by half their consumption and production of CFCs, the principal ODS.[11] But through a series of adjustments and amendments, these targets have been progressively ratcheted up, and now require the complete phase-out of most ozone-depleting substances.[12]

---

4 Convention on International Trade in Endangered Species of Wild Fauna and Flora (1973).

5 Ibid.

6 Protocol of 1978 Relating to the International Convention for the Prevention of Pollution from Ships (With Annexes, Final Act, and International Convention of 1973), Annex I (1978).

7 Canada-United States: Agreement on Air Quality (1991).

8 Convention on Wetlands of International Importance Especially as Waterfowl Habitat, Art 3(1).

9 Ibid.

10 Montreal Protocol on Substances that Deplete the Ozone Layer (1987).

11 Ibid., Art 2.

12 United Nations Environment Programme, 'Handbook for the Montreal Protocol on Substances that Deplete the Ozone Layer' (7th ed 2006).

At the time the climate change issue emerged onto the international agenda in the late 1980s, the ink was barely dry on the Montreal Protocol.[13] Canada hosted the first major international conference on climate change in Toronto in June 1988,[14] less than a year after the Montreal Protocol's adoption, and the U.N. General Assembly adopted its first resolution on climate change that fall.[15] Given the Montreal Protocol's perceived success, many not surprisingly viewed it as a model for the climate change issue and proposed using the same regulatory approach—that is, internationally negotiated, economy-wide targets and timetables.[16] The 1988 Toronto Conference recommended establishing a global emissions reduction target of 20%,[17] and the following year, the Noordwijk Conference recommended that developed countries stabilize their greenhouse gas emissions 'as soon as possible'.[18]

From the start, Western European countries were the biggest proponents of quantitative targets and timetables to limit greenhouse gas emissions, while the United States, Japan, and the Soviet Union voiced skepticism. In Noordwijk, European states were successful in including an emissions stabilization target, but American, Japanese, and Soviet opposition prevented the inclusion of any timetable for achieving that target. Instead, the Noordwijk Declaration merely noted the view of 'many' industrialized

---

[13] For the early history of the U.N. climate change regimes ee generally Daniel Bodansky, 'Prologue to the Climate Change Convention', in Irving Mintzer and J.A. Leonard (eds), *Negotiating Climate Change: The inside Story of the Rio Convention* (Cambridge Univ. Press 1994) 45.

[14] Proceedings of the World Conference on the Changing Atmosphere: Implications for Global Security, June 27–30, 1988, WMO/OMM Doc. 710 (1989), reprinted in (1988) 5 AM. U. J. INT'L L. & POL'Y 515 <http://www.auilr.org/pdf/5/5-2-11.pdf> accessed 05 July 2012.

[15] Protection of Global Climate for Present and Future Generations of Mankind, G.A. Res. 43/53, U.N. GAOR, 43d Sess., Supp. No. 49 (Vol. I), U.N. Doc. A/43/49 (Vol. 1, 6 December 1988) 133.

[16] See, e.g. Scott Barrett, 'International Cooperation and the International Commons' (1999) 10 DUKE ENVTL. L. &POL'Y F. 131, 137; Daniel Bodansky, 'The United Nations Framework Convention on Climate Change: A Commentary' (1993) 18 YALE J. INT'L L. 451, 512; David G. Victor, 'Enforcing Environmental Law: Implications for an Effective Global Warming Regime' (1999) 10 DUKE ENVTL. L. &POL'YF. 147, 160 et seq.

[17] Proceedings of the World Conference on the Changing Atmosphere: Implications for Global Security (n. 13). Although the Toronto Conference was sponsored by the Canadian government, it was non-governmental in character, and the government officials who attended did so in their personal capacity.

[18] Noordwijk Declaration on Atmospheric Pollution and Climate Change, Nov. 7, 1989, 12 Int'l Envtl. Rep. (BNA) 624 (Dec. 13, 1989). In contrast to the Toronto Conference, the Noordwijk Conference was intergovernmental in character.

countries that stabilization should be achieved 'as a first step at the latest by the year 2000'.[19]

The same battle was fought again two years later during the negotiation of the U.N. Framework Convention on Climate Change (UNFCCC),[20] with a similar result. The European Community pushed for the inclusion of targets and timetables to limit greenhouse gas emissions, while the United States argued that states should focus instead on developing national programs and strategies consisting of concrete policy measures.[21] Ultimately, the UNFCCC included aspects of both approaches. Article 4.1 reflects a bottom-up approach, requiring all parties to develop (and report on) national policies and measures to combat climate change—a version of what was referred to, during the negotiations, as 'pledge and review.'[22] Meanwhile, Article 4.2 reflects a top-down model, setting forth a non-binding aim for developed countries to return their emissions to 1990 levels by the year 2000.[23] In essence, the history of the U.N. climate change regime ever since has consisted of variations on these two themes.

2. The Kyoto Protocol: The Ascendance of the Top-Down Approach

The Kyoto Protocol negotiations marked the ascendance, at least temporarily, of the targets-and-timetables approach. In 1995, at the first Conference of the Parties of the UNFCCC (COP-1), states decided to initiate a new round of negotiations aimed at defining quantitative emission limitation and reduction objectives (QELROs) for developed countries—in other words, emissions targets.[24] The negotiating mandate initially left

---

19 Ibid.

20 United Nations Framework Convention on Climate Change, opened for signature 20 June 1992 (entered into force 21 March 1994).

21 Bodansky (n. 15) 468, 514.

22 Ibid., 486 et seq., 508.

23 See generally ibid., 508 et seq. (Interestingly, the bottom-up requirements of Article 4.1 were legally binding, while the top-down target in Article 4.2 was not, illustrating that the issue of top-down vs. bottom-up is distinct from that of legally binding vs. non-legally-binding. Reflecting the principle of common but differentiated responsibilities and respective capabilities (CBDR), the bottom-up requirements of Article 4.1 are common obligations of all parties, while the top-down aim of Article 4.2 is a differentiated provision limited to developed countries.)

24 Conference of the Parties of the United Nations Framework Convention on Climate Change, First Session, Berlin, Ger., Mar. 28–Apr. 7, 1995, Report of the Conference of the

open whether the targets would be legally binding—that is why they were described as 'objectives.'[25] But the following year at COP-2, U.S. negotiators acceded to E.U. demands that the targets under negotiation would be legally binding.[26]

The Kyoto Protocol was adopted in 1987 and establishes emissions limitation targets for each developed country (listed in Annex B), defined as a specified percentage relative to 1990 emissions levels. The targets cover a basket of six greenhouse gases, apply (with only a few limited exceptions)[27] on an economy-wide basis, and cover a five-year commitment period, running from 2008–2012.[28]

The Kyoto Protocol is widely touted for its flexibility. Rather than requiring states to adopt particular policies and measures such as efficiency standards, the Kyoto emissions targets give states freedom in deciding how to reduce emissions and (to a limited degree) where and when to do so.[29] It includes 'flexibility mechanisms' such as emissions trading[30] and the Clean Development Mechanism,[31] which allow states to reduce emissions wherever the reductions are cheapest.[32] And its multi-year commitment period and provision for banking of unused credits[33] give states flexibility about when they reduce emissions.

But although the Kyoto Protocol gives states freedom in how they *implement* their commitments, it does not give them similar flexibility in

Parties on its First Session, Held at Berlin from 28 March to 7 April 1995, 4–6, U.N. Doc. FCCC/CP/1995/7/Add.1 (6 June 1995).

[25] Ibid., 5.

[26] Conference of the Parties of the United Nations Framework Convention on Climate Change, Second Session, Geneva, Switz., July 8–19, 1996, Report of the Conference of the Parties on its Second Session, Held at Geneva from 8 to 19 July 1996, 70, 73, U.N. Doc. FCCC/CP/1996/15/Add.1 (29 October 1996).

[27] The Kyoto targets do not include emissions from international shipping and aviation (usually referred to as 'international bunkers'). In a separate decision adopted in Kyoto, the parties assigned these emissions to the U.N. specialized agencies with substantive competence, the International Maritime Organization for ship-based emissions, and the International Civil Aviation Organization (ICAO) in the case of aircraft emissions. Conference of the Parties of the United Nations Framework Convention on Climate Change, Third Session, Kyoto, Japan, Dec. 1–11, 1997, Report of the Conference of the Parties on its Third Session, Held at Kyoto from 1 to 11 December 1997, 31, U.N. Doc. FCCC/CP/1997/7/Add.1 (25 March 1998).

[28] Kyoto Protocol to the United Nations Framework Convention on Climate Change, Art 3.

[29] Ibid.

[30] Ibid., Art 17.

[31] Ibid., Art 12.

[32] Ibid.

[33] Ibid., Art 3(13).

defining the form and nature of their commitments. Instead, the Protocol prescribes a single type of international commitment (fixed emissions targets, which countries must achieve regardless of changing economic conditions and other national priorities), the scope of those targets (economy-wide), the gases covered (a basket of six greenhouse gases), and the international offsets that can count towards meeting those targets (certified emission reductions created through the collective decision-making procedures of the Clean Development Mechanism).[34] Moreover, although each country has its own national target (E.U. countries, for example, have a –8% target, Japan a –6% target, and Australia a +8% target),[35] these national targets were defined through a process of international negotiations rather than determined by each country unilaterally,[36] and are subject to detailed international accounting rules to determine whether a country has complied.[37]

3. Copenhagen and Cancun: The Bottom-Up Approach Strikes Back

The Kyoto Protocol was a considerable achievement, but it has two significant limitations. First, the states willing to accept Kyoto-style emissions targets represent less than 30% of global greenhouse gas emissions.[38] The United States, which accounts for roughly 25% of global emissions, has refused to join Kyoto, and the Kyoto Protocol—reflecting the principle of common but differentiated responsibilities and respective capabilities—does not establish any emissions limitation commitments for developing

---

[34] Ibid., Arts 3, 12, Annex A.

[35] See ibid., Annex B (setting forth individual national targets).

[36] As a result, under international pressure to get a deal, the United States accepted a minus 7% target in Kyoto, as opposed to the stabilization target that was, in theory, the U.S. 'bottom line' going into the meeting. See ibid.

[37] These accounting rules are mostly set forth in Article 3 and the decisions adopted at the first meeting of the parties on land-use change and forestry, national systems, and the Article 8 review process. Conference of the Parties of the United Nations Framework Convention on Climate Change, Montreal, Can., Nov. 28–Dec. 10, 2005, Report of the Conference of the Parties Serving as the Meeting of the Parties to the Kyoto Protocol on its First Session, Held at Montreal from 28 November to 10 December 2005, U.N. Doc. FCCC/KP/CMP/2005/8/Add.3 (30 March 2006).

[38] These figures are based on the World Resource Institute's Climate Analysis Indicator Tool (CAIT), which compiles figures on global and national emissions. World Resource Institute Climate Analysis Indicator Tool (CAIT) <http://cait.wri.org> accessed 05 July 2012.

countries such as China, which has now surpassed the United States as the world's biggest emitter. Second, the Protocol, in its present form, sets targets for only a five-year commitment period running from 2008–2012. Emissions in 2013 and thereafter are currently unregulated.

Following the adoption of the Kyoto Protocol in 1997, states spent an additional four years negotiating the rules for how the Protocol would work,[39] paving the way for the ratification and eventual entry into force of the Protocol in 2005. But, as soon as this process was completed, attention shifted to the twin problems of what to do after 2012, when the Protocol's first commitment period expires. Should the countries that have Kyoto targets agree to a second round of targets for the post-2012 period,[40] and, if so, should a new agreement be adopted for countries without Kyoto targets, including the United States and China? Or should a single new agreement be adopted that replaces the Kyoto Protocol and is more comprehensive in scope, addressing the emissions of both developed and developing countries?[41] And, more generally, should a post-2012 climate regime continue the top-down architecture of Kyoto or adopt a more bottom-up approach?

In 1995, at the first meeting of the parties to the Kyoto Protocol (CMP-1), states launched a process to negotiate a new round of commitments under the Protocol. Two years later, the UNFCCC parties adopted the Bali Action Plan, which launched a parallel track under the Convention to address the post-2012 period, encompassing all aspects of the climate change issue: mitigation, adaptation, finance, and technology.[42] The

[39] These rules are set forth in Conference of the Parties of the United Nations Framework Convention on Climate Change, Seventh Session, Marrakesh, Morocco, Oct. 29–Nov. 10, 2001, Report of the Conference of the Parties on its Seventh Session, Held at Marrakesh from 29 October to 10 November, 2001, U.N. Doc. FCCC/CP/2001/13 (21 January 2002).

[40] Although Article 3, paragraph 9 of the Protocol—which provides that '[c]ommitments for subsequent periods for Parties included in Annex I shall be established in amendments to Annex B to this Protocol'—could be read as requiring that Annex I countries agree to new commitments for the post-2012 period, it had a more limited purpose, namely, to specify the required method for adopting new commitments (i.e., through amendments to Annex B). Kyoto Protocol to the United Nations Framework Convention on Climate Change, Art 3(9).

[41] See Daniel Bodansky, 'Legal Form of a new Climate Agreement: Avenues and Options' (2009) <www.pewclimate.org/docUploads/legal-form-of-new-climate-agreement-paper.pdf> accessed 05 July 2012.

[42] Conference of the Parties of the United Nations Framework Convention on Climate Change, Thirteenth Session, Bali, Indonesia, Dec. 3–15, 2007, Report of the Conference of the Parties on its Thirteenth Session, Held in Bali from 3 to 15 December, 2007, U.N. Doc. FCCC/CP/2007/6/Add.1 (14 March 2008), 3.

Copenhagen climate conference, held in December 2009, was originally intended as the end point of these parallel negotiating tracks, and many expected it to produce a new legal agreement (or agreements) addressing the post-2012 period—a view reflected in the unofficial slogan of the conference, 'Seal the deal'.[43] The already sky-high expectations were only heightened by the decision by more than one hundred heads of state to attend, including President Obama and the leaders of China, India, Brazil, South Africa, Japan, the United Kingdom, France, and Germany.[44] Thus, when states failed to reach agreement on a new legal instrument, 'Hopenhagen' became seen as 'Nopenhagen' by many disillusioned participants and commentators.[45]

But the Copenhagen meeting was far from the flop often portrayed. Although it failed to adopt a legally binding treaty, it did produce the Copenhagen Accord, a political agreement negotiated by the leaders of more than twenty-five countries in the closing hours of the meeting.[46] Inelegant and extremely brief, the Copenhagen Accord nevertheless addresses all of the main elements under negotiation, including mitigation, adaptation, finance, technology, forestry, and verification. Among its key elements, it:

- sets a long-term aspirational goal of limiting temperature rise to no more than 2°C;
- establishes a process for recording the mitigation targets and actions to be implemented by developed and developing countries (which the principal countries had put forward prior to the meeting);
- puts significant new money on the table for climate change mitigation and adaptation by developing countries, including 'fast-start' money for the 2010–2012 period 'approaching' $30 billion and a goal of mobilizing $100 billion per year by 2020; and

---

[43] Richard Black, 'Ban Ki-moon Tells Copenhagen Summit to "Seal a Deal"', *BBC News* (last updated 15 December 2009) <http://news.bbc.co.uk/2/hi/science/nature/8414798.stm> accessed 05 July 2012.

[44] The U.N. Climate Change Conference in Copenhagen, Agenda, UNFCCC Webcast, <http://cop15.meta-fusion.com/kongresse/cop15_hls/templ/ovw_copenhagen.php?id_kongressmain=101> accessed 05 July 2012.

[45] See generally Daniel Bodansky, 'The Copenhagen Climate Change Conference: A Postmortem' (2010) 104 Am. J. Int'l L. 230.

[46] Conference of the Parties of the United Nations Framework Convention on Climate Change, Fifteenth Session, Copenhagen, Denmark, Dec. 7–19, 2007, Report of the Conference of the Parties on its Fifteenth Session, Held in Copenhagen from 7 to 19 December 2009, U.N. Doc. FCCC/CP/2009/11/Add.1 (30 March 2010) 4 et seq.

- provides for 'international consultation and analysis' of all developing country mitigation actions, plus fuller monitoring, reporting, and verification (MRV) of those actions that receive international support and of developed country targets and financing.[47]

The Copenhagen Accord embraces a fundamentally different architecture than the Kyoto Protocol. Rather than defining emissions targets from the top down through international negotiations, the Copenhagen Accord establishes a bottom-up process that allows each party to define its own commitments and actions unilaterally.[48] The Accord specifies that developed countries will put forward national emissions targets in the 2020 time frame, but allows each party to determine its own target level, base year, and accounting rules.[49] Meanwhile, developing countries have even greater latitude in formulating nationally appropriate mitigation actions (NAMAs).[50]

Of course, in accepting the Copenhagen Accord, states already knew each other's pledges, since these had been announced prior to the meeting. And there was a strong expectation that states would stick by these pre-meeting pledges when it came time to recording their targets and actions internationally—an expectation that was, in fact, fulfilled. But, consistent with the bottom-up approach, the Copenhagen Accord simply accepted states' pledges, rather than subjecting them to international negotiations in an effort to develop stronger commitments, as had been the case in Kyoto.[51]

Although the Copenhagen Accord suggests a significant reorientation of the climate change regime, its status coming out of Copenhagen was unclear. On the one hand, it had been adopted by the leaders of all of the world's major economies, giving it considerable weight.[52] On the other hand, when it was brought back to the formal Conference of the Parties in the waning hours of the meeting, the Conference could agree only to 'take note' of it, rather than to adopt it, due to objections from a handful of

---

47 Ibid.
48 Ibid.
49 Ibid., 6.
50 Ibid., 4, 5.
51 See ibid., 4 (requiring states to 'submit' their quantified emissions targets for 'compilation in an INF document').
52 See ibid., 5 (listing the delegations agreeing to the Accord).

countries led by Sudan, Venezuela, and Bolivia.[53] As a result, the Accord had no official status within the UNFCCC process and some reportedly argued that the UNFCCC secretariat should not even record the national pledges that were made in spring 2010 on its official website.[54]

Given this situation, the decisions adopted in December 2010 at the Cancun Conference were an impressive accomplishment. The Cancun Agreements not only brought the various elements of the Copenhagen Accord into the UNFCCC process, but elaborated the Accord's three-page text into thirty pages of decision language.[55] Key elements of the decisions include:

- a reiteration of the long-term goal of limiting temperature increase to 2°C;
- anchoring of the emissions targets and actions pledged pursuant to the Copenhagen Accord in the UNFCCC process, through inclusion in two 'INF' (information) documents—one for emissions targets to be implemented by developed countries, the other for NAMAs to be implemented by developing countries;
- establishment of a registry for listing NAMAs for which developing countries are seeking international support;
- establishment of the Green Climate Fund, which will be managed by a 24-member board of directors and administered for the first three years by the World Bank;
- reiteration of the collective commitment in Copenhagen by developed countries to provide an amount approaching $30 billion in fast-start financing for the 2010–2013 period, balanced between mitigation and adaptation, as well as of the longer-term goal of mobilizing $100 billion per year by 2020, a 'significant portion' of which should flow through the newly established Green Climate Fund;

---

53 See ibid. (not including Sudan, Venezuela, and Bolivia in the list of delegations agreeing to the Accord).

54 Jacob Werksman, 'Associating' with the Copenhagen Accord: What Does it Mean?' *World Resources Institute* (25 March 2010) <http://www.wri.org/stories/2010/03/associating-copenhagen-accord-what-does-it-mean> accessed 05 July 2012.

55 Conference of the Parties of the United Nations Framework Convention on Climate Change, Sixteenth Session, Cancun, Mexico, Nov. 29–Dec. 10, 2010, Report of the Conference of the Parties on its Sixteenth Session, Held in Cancun from 29 November to 10 December 2010, U.N. Doc. FCCC/CP/2010/7/Add.1 (15 March 2011).

- elaboration of the process of international consultation and analysis of developing country mitigation actions, including that it will be performed by the Subsidiary Body on Implementation (SBI) of the UNFCCC;
- establishment of a new technology mechanism to facilitate technology development and transfer;
- establishment of a framework for reducing emissions from deforestation and forest degradation (REDD); and
- adoption of the Cancun Adaptation Framework.[56]

Some commentators, in comparing Copenhagen and Cancun, have contrasted the 'Chihuahua that roared' with the 'Great Dane that whimpered'.[57] But these differing assessments reflect differences not so much in the outcomes of the two meetings than in the advance expectations. Many expected Copenhagen to adopt a new set of legally binding emissions targets, á la Kyoto, so the nonbinding Copenhagen Accord was a major disappointment. In contrast, expectations for Cancun were rock-bottom, so an outcome that reiterated and elaborated the Copenhagen Accord was seen as a big win.

This is not to diminish the significance of the Cancun conference. Success was uncertain until the very last, and was a testament to the skill of the Mexican chair, Foreign Minister Patricia Espinosa, who did a masterful job both in creating a negotiating atmosphere in which countries had confidence, and in putting pressure on Venezuela and its allies not to torpedo the meeting. In the end, when only Bolivia objected to the adoption of the text, Espinosa was able simply to note the Bolivian objection and then gavel through the decision to widespread applause.[58]

4. Comparing the Top-Down and the Bottom-Up Approaches

Although negotiations under both the Kyoto and Convention tracks are continuing to proceed (thus leaving open the possibility of a new

---

[56] Ibid.

[57] Richard Black, 'The Chihuahua That Roared' *BBC News* (11 December 2010) <http://www.bbc.co.uk/blogs/thereporters/richardblack/2010/12/cancun_the_chihuahua_that_roar.html> accessed 05 July 2012.

[58] The Cancun outcome was also helped by two other factors: first, a widespread fear that, if Cancun failed, this would discredit the entire UNFCCC process, and second, the fact that states had had a year to digest the Copenhagen Accord, as opposed to the few minutes they were given in Copenhagen itself to decide whether to adopt the Accord. See ibid.

top-down agreement that extends, complements, or replaces the Kyoto Protocol), the Cancun Agreements legitimized for now the bottom-up process launched by the Copenhagen Accord. Thus far, more than eighty countries, including all of the world's major economies, have made specific national mitigation pledges, which have been recorded by the UNFCCC Secretariat.[59]

Will the bottom-up approach reflected in the Copenhagen/Cancun agreements prove sufficient, either as is or as part of a new legal agreement? Or is a top-down, target-based approach like the Kyoto Protocol needed to address the climate change problem?

In thinking about this issue, it is useful initially to distinguish it from the related issue of legal form: whether whatever architecture is adopted will be reflected in a political agreement like the Copenhagen Accord or in a legally binding treaty like the Kyoto Protocol. As the UNFCCC illustrates, a bottom-up architecture can be embodied in a treaty and be legally binding—that is the essence of Article 4.1 of the Convention, which requires countries to develop policies and measures to address climate change. The Obama administration supports this kind of approach.[60] Conversely, a top-down target can be nonlegal in character, either because it is contained in a political agreement or because it is phrased in aspirational rather than mandatory terms (as was true of the target set forth in Article 4.2 of the Convention).

Proponents of the top-down approach argue that, absent internationally negotiated emission targets, states will do very little to combat climate change.[61] As they note, climate change is a classic example of a collective-action problem, which can be solved only through cooperative action by all of the world's major emitters.[62] Individual states have little incentive to act unless they are confident that their actions will be reciprocated by others; otherwise, they will simply incur economic costs without getting

---

[59] Communications Received from Parties in Relation to the Listing in the Chapeau of the Copenhagen Accord <http://unfccc.int/meetings/cop_15/copenhagen_accord/items/5276.php> accessed 05 July 2012.

[60] See, e.g., Ad Hoc Working Group on Long-Term Cooperative Action Under the Convention, Sixth Session, Bonn, Ger. June 1–12, 2009, Ideas and Proposals on the Elements Contained in Paragraph 1 of the Bali Action Plan, U.N. Doc. FCCC/AWGLCA/2009/Misc.4 (Part II), (19 May 2009) 106 (hereinafter U.S. Submission on Copenhagen Agreed Outcome).

[61] William Hare et al., 'The Architecture of the Global Climate Regime: A Top-Down Perspective' (2010) 10 Climate Pol'y 600, 604 et seq.

[62] On collective action problems, see generally Scott Barett, *Environment and Statecraft: The strategy of Environmental Treaty Making* (Oxford Univ. Press 2003).

much environmental benefit. The task of the climate change regime is to ensure reciprocity of effort, by establishing legally binding commitments for each party and providing some assurance of compliance.

Although this collective action analysis seems correct in theory, it has not done very well in practice in accounting for the behavior of key actors. On the one hand, some actors have pushed forward to reduce their emissions, without waiting for their efforts to be reciprocated by others. In the United States, some states, cities, and businesses have developed their own climate policies, rather than waiting for national or international action.[63] Similarly, the European Union has forged ahead with its emissions trading scheme even though the United States and China lack national emissions targets.[64] And the E.U. seems committed to continuing down this path whether or not there is any international agreement that extends or replaces Kyoto.[65]

On the other hand, developing countries seem unwilling to accept emissions targets defined through international negotiations, even as part of a global deal involving reciprocal actions by all of the world's major emitters. They seem to prefer no deal at all to a deal that would involve their accepting internationally-defined limits on their emissions. And although the Obama Administration professes to support a legally-binding agreement on climate change,[66] it is doubtful that the U.S. Senate would consent to *any* new climate change treaty, much less one with internationally-defined emissions targets, even if it had comparable targets for developing countries. Moreover, even the Obama Administration has proposed making national emissions targets subject to national accounting rules, thereby allowing national law to play a role in defining any U.S. emissions target.[67]

---

[63] For information on state and local efforts within the United States, see Pew Center on global Climate Change <www.pewclimate.org/states-regions> accessed 05 July 2012.

[64] See European Commission, Climate Action: Emissions Trading System <http://ec.europa.eu/clima/policies/ets/index_en.htm> (last updated 15 November 2010) accessed 05 July 2012.

[65] The European Commission has already announced plans for Phase III of the EU Emissions Trading System (covering the period after 2013), which are not contingent on adoption of an international agreement for the post-2012 period. See EU ETS Phase III (2013–2020), UK Department of Energy and Climate Change <www.decc.gov.uk/en/content/cms/emissions/eu_ets/phase_iii/phase_iii.aspx> accessed 05 July 2012.

[66] See Todd Stern on the U.S. Support to a Legally Binding Agreement Including All Major Economies, *YouTube* (16 May 2011) <www.youtube.com/watch?v=zcmfPwDOkuo> accessed 05 July 2012.

[67] The proposed U.S. agreement would allow states to implement their international commitments 'in conformity with domestic law'. See U.S. Submission on Copenhagen

What accounts for the reluctance of many states to accept top-down, Kyoto-style targets, even as part of a global deal that required reciprocal actions by others? Perhaps the main reason is that, in many countries, climate change is seen more through a domestic prism than an international one. Climate change implicates virtually every area of domestic policy, including industrial, agricultural, energy, transportation, and land-use policy. Hence, it raises huge domestic sensitivities—much more so than any previous environmental issue. Building a domestic coalition to address the problem is enormously complicated as it is, and would be made even more difficult by internationally-negotiated requirements that constrain a state's flexibility. Simply put, for many countries, the costs of emissions targets to national sovereignty outweigh their benefits in ensuring reciprocal emissions reductions by other countries.

A bottom-up approach, in contrast, allows each state to go at its own pace, in its own way. International pledges grow out of, and reflect, domestic policies, rather than being superimposed on them. The role of the international regime is not to define what each state must do, but rather to help generate greater political will by raising the profile of the climate change issue and providing greater transparency.

Critics contend that national pledges will be weak, and will not provide enough emissions reductions to prevent dangerous climate change.[68] The national pledges made pursuant to the Copenhagen Accord seem to bear out this fear. As several analyses indicate, they do not, in fact, put the world on a realistic pathway towards limiting climate change to 2°C, the professed goal of the Accord.[69]

The question, however, is not whether the Copenhagen pledges are adequate—clearly they are not—but rather whether a Kyoto-style agreement would do better. According to collective action theory, states should be willing to do more to combat climate change as part of a legal agreement that provides for reciprocal actions by all of the major emitters. But,

---

Agreed Outcome (n. 59) 107. Although the meaning of this language is not altogether clear, it appears to allow countries, through their national legislation, to specify their targets in different ways—for example, with respect to sectoral coverage, base years, or allowable offsets.

68 See Hare et al. (n. 60) 601.

69 See UNEP, 'The Emissions Gap Report: Are the Copenhagen Accord Pledges sufficient to limit Global Warming to 2°C or 1.5°C? A preliminary Assessment' (2010) 46 <www.unep.org/publications/ebooks/emissionsgapreport/pdfs/EMISSION_GAP_REPORT_LOWRES.pdf> accessed 05 July 2012; Joeri Rogelj et al., 'Copenhagen Accord Pledges are Paltry' (2010) 464 NATURE 1126, 1126 et seq.

apart from the European Union, which has pledged to boost its emissions reduction target from 20% to 30% as part of a global deal,[70] there is little indication that countries feel this way.

5. Conclusion

Everyone wants to learn from history, so as not to repeat it. But what are the lessons of the Kyoto Protocol? Although opinions differ widely, I believe it suggests the need for a more evolutionary approach to the development of the climate change regime.[71]

Most international regimes do not emerge all at once; they develop gradually. In some cases, regimes start with a relatively small group of countries and, over time, become broader; in others, they start relatively weak and, over time, become stronger. And, in some cases, they do both—witness the European human rights system and the international trade regime.

In the long run, unless a technological magic bullet can be found, solving the climate change problem may well require a collective agreement among the key contributors, as proponents of the top-down approach claim. But the Kyoto experience suggests that many states are not yet ready for this solution.

If this assessment is correct, then the climate change regime needs to proceed in a more incremental way, to provide time for countries to learn from experience and to develop trust in the system. The Copenhagen/Cancun agreements establish a process for countries to develop and report on national actions to reduce emissions, and for the international community to review what countries are doing. The agreements also call for a more general review, to be concluded in 2015, to consider the adequacy of the 2^0 goal and the regime's overall progress in achieving it. If countries' initial experience under the agreements is positive, this could encourage them to strengthen the regime—for example, through more detailed

[70] The EU has pledged to reduce its emissions unilaterally by 20% by 2020. But it has indicated that it would reduce its emissions by 30% as part of a global deal involving reciprocal actions by the other major emitters. For information on European Union climate change policies, see European Commission, 'Climate Action: What is the EU Doing on Climate Change?' <http://ec.europa.eu/clima/policies/brief/eu/index_en.htm> (last updated 18 October 2010) accessed 05 July 2012.

[71] See generally Daniel Bodansky and Elliot Diringer, 'The Evolution of International Regimes: Implications for Climate Change' (2010) <www.pewclimate.org/docUploads/evolution-multilateral-regimes-implications-climate-change.pdf> accessed 05 July 2012 (concluding that the evolutionary process is politically the most promising way forward).

requirements for national pledges or stronger reporting and review procedures, or by incorporating the Copenhagen/Cancun architecture into a legally-binding agreement.

The comparatively weak provisions of the Copenhagen Accord and Cancun Agreements are unlikely, in themselves, to produce the necessary level of emissions cuts to prevent dangerous climate change. For this reason, they need to be complemented by actions in other arenas. The Montreal Protocol, for example, has already made a huge contribution to the climate change issue, by accelerating the phase-out of HCFCs,[72] and many states support using it to regulate HFCs, which are currently regulated by the Kyoto Protocol. But the Copenhagen/Cancun agreements represent an important step forward. They have unblocked what appeared to be a stalemated process, and help build a foundation for stronger action in the future.

[72] See Guus J.M. Velders et al., 'The Importance of the Montreal Protocol in Protecting Climate' (2007) 104 PROC. NAT'L ACAD. SCI. U.S. 4814, 4817; Donald Kaniaru (ed), *The Montral Protocol: Celebrating 20 Years of Environmental Progress—Ozone Layer and Climate Protection* (Cameron May 2007).

ENERGY SWITCH IN GERMANY: 100% RENEWABLE ELECTRICITY BY 2050

Heidi Foth[1] and Sönke Bohm[2]

1. Introduction

The future of Germany's and Europe's energy supply is currently the subject of a broad debate where energy policies are inextricably bound up with environmental and climate policies. In order to avert potentially catastrophic climate change by limiting global temperatures to 2 degrees Celsius above pre-industrial levels, industrialised countries such as Germany need to reduce their carbon emissions by 80–95 percent by 2050. In order to meet these needs, German power plants would need to be virtually emission free because emissions cannot be reduced sufficiently by 2050 in other sectors for technical reasons.

Germany's energy supply system needs to face a transition with a view to achieve by 2050 an energy supply that is economically sustainable, reliable, and climate-friendly. The special report 'Pathways towards a 100% renewable electricity system' of the Advisory Council on the Environment (SRU)[3] discusses the need for transformation of our electricity supply system, in light of the relevant technical, economic, legal and political factors, and with the goal of ensuring that Germany is able to establish a sustainable and decarbonized electricity supply system by 2050.

A considerable share of Germany's current electricity generation capacities will need to be replaced over the next two decades, since many power plants will have reached the end of their lifetime by then. The investments that are made in the coming years will have a major impact on both the structure and emissions of the electricity sector for decades to come. This situation presents an opportunity to revamp Germany's power plant fleet in a manner that will be relatively inexpensive and that will constitute a far reaching structural change in our country's electricity

[1] Institute of Environmental Toxicology, Martin Luther University, Franzosenweg 1a, Halle Saale, e-mail: heidi.foth@medizin.uni-halle.de.

[2] International Institute for Management, University Flensburg, Munketoft 3b, 24937 Flensburg, e-mail: soenke.bohm@uni-flensburg.de.

[3] See <www.umweltrat.de> accessed 11 July 2012.

system. In order to meet these needs, German power plants would need to be virtually emission free because emissions cannot be reduced sufficiently by 2050 in other sectors (such as agriculture and goods) transport for technical reasons.

The SRU has laid out the reasons why a sustainable and climate-friendly electricity supply can only be achieved using renewables and this opinion is in line with a series of recent studies that have investigated whether a largely or wholly renewables based electricity supply is achievable in Germany and Europe. The decisive issues are addressed, whether the transition is technically feasible, what are the expected costs and timelines. Moreover, other issues include which hurdles need to be cleared and which political and legal frameworks need to be taken into account. The various scenarios that would allow for establishment of a wholly renewable electricity supply in Germany by 2050 are described. The political hurdles for transitioning to a wholly renewable electricity supply with a strong European perspective, are discussed in particular: the new EU energy and environmental policy competency framework pursuant to the Lisbon Treaty; the future of EU climate and energy policy; bilateral and multilateral cooperation; the policy conditions needed nationally to transition to a wholly renewable electricity supply.

2. Challenges to Meet Energy Supply

Greenhouse gas emissions in industrialised countries need to be reduced by 80 to 95 per cent by 2050 in order to avert a dangerous rise in global temperature amounting to more than 2°C relative to the pre-industrial level. This goal was officially endorsed by the European Council in 2009. In Germany, there is a broad political consensus about ambitious climate protection objectives and the Federal Government has recognized the need to reduce greenhouse gases by at least 80 percent by 2050.

In its current form, electricity accounts for roughly 20 percent of energy demand but for 40 percent of total German carbon emissions, thus making electricity supply a key area of activity in Germany's energy and climate policies. If carbon emissions are to be reduced by 80 to 95 percent in Germany, German power plants would need to be virtually emission-free. This is due to the fact that, for technical reasons, in other sectors such as agriculture and goods transport emissions cannot be reduced sufficiently by 2050, or the costs of such reductions would be prohibitive. To electricity companies the requisite technological solutions are already available.

Germany will be facing key decisions in the coming years concerning the structure of its power supply. Much of its generation capacity will need to be replaced over the next two decades since many conventional power plants will be nearing retirement by then. The investments that are made in the coming years will have a major impact not only on the structure but also on the emissions associated with the electricity sector for decades. This situation presents an opportunity to set in motion a relatively low-cost but far-reaching structural change of the sector. A sustainable and climate friendly electricity supply system can—over the long term—only be achieved if it is based on renewable energy sources. Such a goal represents an opportunity for sustainable technological and infrastructure innovation. The special report 'Pathways towards a 100% renewable electricity system' of the Advisory Council on the Environment SRU discusses the need for transformation of our electricity supply system, in light of the relevant technical, economic, legal and political factors, and with the goal of ensuring that Germany is able to establish a sustainable and decarbonized electricity supply system by 2050. The SRU's special report is one in a series of recent studies that show that in Germany and Europe as a whole a complete or nearly complete shift of the power supply to renewables could be achieved.

3. Point of Departure in the Discussion

In September 2010, the German government issued a scenario-based energy concept containing a far-reaching electricity supply roadmap. Many elements of the government's energy concept, particularly those relating to the period to 2050, the long-term objectives laid down in the concept, and the proposed electricity grid expansion measures could potentially be regarded as exemplary at the international level. That said, the SRU's special report goes further than the government's energy concept in terms of both climate protection objectives and the envisaged energy mix, while at the same time completely parting ways with the government's strategies for the transition to renewables by 2050. In the SRU's view, the prospects for this transition are far brighter than the government presumes. On the other hand, the SRU is far less persuaded than the government appears to be concerning the compatibility of nuclear power and renewables. But many of the recommendations and concepts in the special report are relevant regardless of whether the goal is to achieve 80 or 100 percent renewable electricity.

In decisions about technology, energy policymakers should consider not just on how climate-friendly a particular technology happens to be, but also the overall statutory framework for environmental sustainability at the international level (Rio Declaration, UNFCCC), EU level, and national level (article 20a of the German Constitution; the German government's sustainability strategy). The main sustainability factors that come into play in this regard are as follows: compliance with the absolute sustainability and input limits of natural systems; and taking steps to ensure intergenerational and global justice (instituting equal per-capita usage levels for common-pool resources). The Earth's climate system and biodiversity are natural systems whose ecological input limits have already been exceeded and that are urgently in need of protection. In order to meet the aforementioned sustainability criteria in particular (maintaining the sustainability of natural systems; achieving generational equality), it will be of course necessary to minimise the risk of irreversible or catastrophic events.

4. Sustainability Assessment of Technology Options

Transitioning towards a low carbon power sector is currently at the centre of the political debate and is also a key issue for the SRU. The low carbon technologies needed to do this are already available, or will be in the foreseeable future. These technologies include renewable energy such as wind, solar, biomass and geothermal energy, nuclear power, and fossil fuel power generation using carbon capture and storage (CCS) technology. Sustainability is more than just reducing greenhouse gases by lowering emissions of carbon dioxide. It is also an interactive network of different factors and processes; some of which are in competition to each other. The sustainability criteria to address in the decision are: Lifetime, land use, site selection, emissions, climate protection, safety acceptance, disposal, reversibility, compatibility and integration. The following paragraph will touch on some issues for the technical options of electricity production.

In view of the fact that all energy generation technologies affect the eco-balance in one way or another, there is simply no such thing as 100 percent environmentally neutral energy generation. However in the overall balance, the comparative sustainability assessment (carried out by the SRU) shows that renewables constitute the only sustainable energy option. The main goal that needs to be aimed at—fully decarbonised electricity generation—cannot be reached either through more efficient conventional coal fired power plants or carbon capture and storage technology. Moreover, the use of coal fired power plants entails large-scale raw material extraction operations which despite improved air purity efforts result in significant air quality problems. As for CCS, its use is limited by the available storage capacity and competition from other potential uses of this capacity. Although greenhouse gas emissions from nuclear power plants are far lower than for coal fired power plants, the use of nuclear power entails the risk of accidents. This is an eventuality that cannot be completely ruled out and that could have consequences for large areas and for extended periods of time. Moreover, no viable solution has been found for long-term storage of nuclear waste. In the SRU's view, this is a high price to pay. Moreover, nuclear power may not be a sustainable solution in view of the limited supply of uranium. Hence the SRU believes neither coal fired power plants nor nuclear power can qualify as sustainable energy resources.

But renewables are not always without problems either. For instance, growing energy crops may provoke land use changes that have a substantially negative climate warming effect, they may have a deleterious impact on natural capital, they may cause considerable environmental damage, and they may run afoul of the principle of equal usage rights by competing with food crop production. Other renewables and the transmission and storage capacity expansion needed for them can also provoke conflicts in terms of land or ocean use. Moreover, renewables entail the use of resources such as water and rare metals. Altogether, the weight of impact from the ecological problems associated with renewables are manageable and could be minimised through policy and planning measures. The environmental problems posed by coal fired power plants and nuclear power plants are mainly of a technological nature and the same, regardless of location, whereas renewables provide some leeway in terms of localization. Furthermore, actual energy consumption is far lower than the putative potential offered by renewables. Ecological conflicts could be largely mitigated if power plant construction was supported by regional planning.

Another key factor here is that renewable-energy power plants are normally smaller and easier to dismantle than the counterpart conventional facilities. Thus they are more flexible infrastructure components whereas nuclear and CCS coal fired power plants are associated with long-term consequential environmental damage and risk resulting from coal mining, nuclear waste storage, and carbon storage. The environmental impact of renewable energy is generally confined to the service life of the installation. Furthermore, at least for solar and wind power, their environmental impact is confined to the plant construction phase, whereas fossil and nuclear power plants necessitate sustained land and natural resource use to mine the fuels needed. Renewables are not only sustainable, but are also in line with the precautionary principle. Therefore, in view of the present uncertainties, they can be adapted more flexibly to changing conditions and have a greater tolerance for error. Hence renewables are superior to conventional energy resources in terms of generational equity and risk avoidance, thus making renewables more sustainable. This in turn means that renewables are the only viable sustainable solution for electricity generation.

4.1. *The Goal: Sustainable Electricity Supply in 2050*

The crucial questions are whether a wholly renewables-based electricity supply is technically feasible for and in Germany and whether such a system will ensure security of energy supply that is on a par with today's energy supply. The problem was approached by computational modelling of different scenarios of energy demand, transnational networking, the potential of renewable energy sources such as hydropower, wind on/off shore, biomass, photovoltaic and geothermal energy and learning curves regarding technology, production, costs, energy efficiency and economies of scale. The scenarios analysed by the SRU and laid out in the SRU's special report show that there are various options for the institution of a wholly renewable electricity supply in Germany.

These scenarios were developed at the SRU's request by the German Aerospace Centre (Deutsches Zentrum für Luft- und Raumfahrt, DLR) using the DLR's REMix model. The REMix model is based on a geoinformation system which, using a high resolution grid, documents the electricity generation potential of all renewable energy sources in Germany, Europe and North Africa, and then uses the results to compute a cost optimised (i.e. lowest possible cost) energy portfolio for the defined conditions. In as much as the model uses one hour time intervals, it can correlate annual electricity generation with electricity demand down to the hour. This in turn meets the challenges posed by an electricity system that makes increased use of wind and solar energy, whose availability varies over time. A renewables-based electricity supply that satisfies demand at all times must be achieved either through the use of overlapping electricity production from renewable resources and/or stored electricity. To this end, for the target year 2050 the DLR modelled the use of hydropower, in conjunction with wind, solar, biomass and geothermal energy as well as electricity storage technologies such as pumped hydro and compressed air energy storage and developed cost-optimised constellations of these energy technologies for each relevant instance. These calculations presupposed that learning curve effects will drive down the cost of renewable energy technologies and hence renewably generated electricity over the long term. The cost suppositions used by the DLR in the REMix model are the fruit of thorough research and continuous updating, but are regarded in some quarters as overly optimistic and in others as unduly pessimistic. Hence, the actual costs associated with renewable energy technologies going forward may turn out to be higher or lower than those predicted in the special report, and the actual cost-optimised portfolio of renewable energy sources may differ from the structure that was modelled.

The SRU's scenarios are confined to those renewable energy technologies that are already well established and for which a reasonably reliable estimate of future costs can be made. Although other renewable energy technologies such as wave and tidal energy are in the pipeline offering additional options and leeway for implementation of a fully renewable electricity supply, these technologies were excluded from the DLR model owing to a lack of reliable data. All eight of the scenarios the SRU presents in the special report presuppose that Germany can and will implement a wholly renewable electricity supply by 2050, albeit under varying conditions in respect to grid connections with other countries and the electricity demand that will need to be met. The SRU also compares the following putative models with each other:

- German energy self-sufficiency (scenario group 1);
- a regional network involving Germany and Scandinavia (scenario group 2); and
- a Europe-North African network (scenario group 3).

Scenario group 1 can be regarded as a last-resort worst case scenario. It is not a desirable solution for economic reasons or from the standpoint of EU law. It can be shown, though, that even a comparably high estimate of electricity demand amounting to 700 TWh in 2050 can be met with domestic renewables alone. It follows that things will be much easier under less restrictive conditions. The other scenarios presuppose that all participating countries will jointly seek to transition to renewables. For reasons of security of supply, each such country has a vested interest in generating a maximum percentage of its electricity from domestic resources. A maximum net import of 15 percent of total generated electricity was predicated.

All of the scenario computations are based on projected total electricity demand amounting to 500 and 700 TWh, respectively, in Germany. The SRU presumes that a German electricity demand of 500 TWh is achievable over the long term, even if electricity demand rises substantially in the heating and transportation sectors. But to do this, it will be necessary to leverage the efficiency optimisation potential not only for electricity use but also in the areas of building renovation, heating and hot water provisioning. The scenarios based on electricity demand amounting to 700 TWh clearly show that a wholly renewable electricity supply would be within reach from a technical point of view even if energy efficiency and savings policies are not successfully implemented and energy consumption is above current demands.

However, as is always the case with long range scenario studies, the findings presented are subject to significant uncertainty as it was necessary to make a series of suppositions concerning potential developments in the whole process that are difficult to forecast. The scenarios are intended to show that a wholly renewable electricity supply is within reach under various conditions. They do not constitute a forecast of the potential developments that might come into play here and they do not indicate a preference on the part of the SRU for a specific portfolio of renewables or for any particular cross-border electricity supply network solution. Additionally, they are not intended as a blueprint for the transition to a wholly renewable electricity supply. Instead, the scenarios merely provide a selection of the many possible solutions that come into play here.

4.2. *The Path to Transition*

As mentioned earlier, the envisaged upgrading of Germany's power plant fleet offers a golden opportunity to transition to a wholly renewable electricity supply at a relatively low cost and without engendering any discontinuities in supply structures. In view of the fact that ambitious energy saving and efficiency policies would go a long way toward easing the transition to a wholly renewable electricity supply, efforts should be made to reduce and—over the long term—stabilise Germany's electricity demand. This would lower the economic and ecological costs of the system, improve its robustness, and promote rapid implementation of the necessary transformation process.

As a second step, the findings from the 2050 scenarios were used to determine how the available electricity generation, transmission and storage capacities would need to evolve in order to achieve the defined target state by 2050. Of particular interest was to elucidate:

1. How much a wholly renewable electricity supply and the associated transition will probably cost.
2. What would be a realistic timeline for the transition to such an electricity supply and which measures would this transition entail?

Based on the characteristics of the existing power plant fleet, the SRU shows how conventional generation capacity could be replaced incrementally by renewable energy generation capacities.

The scenario computations show that Germany could readily achieve a wholly renewable electricity supply that is both reliable and affordable. Providing that the relevant storage facilities and grids are implemented, the renewable energy potential in Germany and Europe would allow for the satisfaction of maximum projected electricity demand at all times throughout the year, using wind turbines, solar collectors (pv. and/or solar thermal power plants), and other currently available technologies and despite fluctuations in the availability of renewable electricity. As the lowest cost energy resource in the run-up to 2050, wind energy, particularly from offshore wind turbines, plays a pivotal role in all of the scenarios discussed in the special report. On the other hand, the level of solar energy use in the various scenarios varies according to electricity demand and the amount of electricity that is imported. Biomass use in the scenarios involving transnational power supply networks accounts for no more than 7% of electricity demand, largely owing to land use conflicts and the relatively high cost of this energy resource.

Although a wholly renewable domestic electricity supply without any electricity imports would be feasible, this option should definitely not be pursued in light of the evolving EU-wide internal market for energy. Although various European inter-regional networks may well be feasible, the SRU selected only two of such scenarios to illustrate the feasibility. A Germany-Denmark-Norway power network would allow for the use of the enormous pump storage system potential in Scandinavia. This option, which has been largely neglected in the energy policy debate in Germany in the past, has been analysed in greater detail. That said, a more far-flung European power supply network with a number of EU states availing themselves of low-cost response power (particularly from Norway) would help the states achieve a wholly renewable electricity supply. In any case, this can only be achieved by expanding offshore wind power capacity and high capacity lines in the North Sea region, and through the use of pump storage systems.

In the view of the SRU, instituting a wholly renewable electricity supply in Germany by 2050 would entail economic advantages in addition to promoting climate protection. The aggregate costs of such a system would be largely determined by the extent to which a network comprising other European countries is established. Inflation adjusted, a wholly renewable electricity supply using German resources only would be relatively cost intensive, ranging from 9 to 12 Euro-Cents per kWh, depending on the demand size. On the other hand, an inter-regional smaller-scale German-Danish-Norwegian or larger-scale Europe-North Africa network would provide electricity at a cost of only 6 to 7 Euro-Cents per kWh, including the cost of international grid and storage capacity expansion. Rough estimates indicate that expanding the German grid would entail additional costs amounting to approximately 1 to 2 Euro-Cents per kWh.

Over the long term, renewable electricity will prove to be less cost-intensive than conventional low carbon technologies such as CCS power plants and new nuclear power plants whose costs will rise owing to limited uranium resources and storage facilities respectively. The price of nuclear energy will also be driven upward by the currently unforeseeable costs of long-term nuclear waste storage, whereas renewable energy costs will decline owing to the effects of learning curves and economies of scale. Whereas timely short-term expansion of renewable energy resources will entail higher near-term investment and generation costs than the cost of extending the service life of existing power plants, it will nonetheless allow for long-term cost savings not only in terms of direct costs but even more so in terms of social costs, and is thus a worthwhile investment in the future. The SRU estimates that such savings from a wholly renewable electricity supply will begin to be seen between 2030 and 2040 if the attendant costs decrease as described in the special report, and somewhat later if these costs decline more slowly than forecast.

4.3. *The Elements of Transition*

When it comes to determining the way forward for implementation of an electricity system that will make a major contribution in the fight against global warming, the government needs to opt for solutions that are maximally compatible with the sustainability and precautionary principles laid down in Article 20a of the German Constitution. In the SRU's special report, the case has been made that an energy policy that promotes anything other than renewable energy sources over the long term is at odds with the stipulations of the aforementioned constitutional clause. Transitioning to a renewables-based electricity supply needs to be widely accepted by the general public and consistent with the classic energy policy goal of assuring a reliable supply of affordable energy. In view of this key exigency, the special report assesses the technical, economic and political feasibility of instituting a fully renewables-based electricity system in Germany.

Political measures and instruments of policy control are needed to bring about this transformation smoothly and efficiently. So the question is posed: which political and legal frameworks would need to be taken into account for such a transition in the European context and how much leeway do they allow?

The starting points are that deficiencies and problems prevail concerning the lack of nationwide network planning. The search for routes is yet informal and preliminary and regional planning does not cover the needs. The SRU report suggests a two stage planning with environmental impact assessments (SUP/UVP-FFH) with the following elements:

- National transmission networks plan;
- Planning process with final decision of the government;
- Allocation of administrative competences to federal bodies.

The intended consequences should be clear procedural steps, elimination of informal preliminary decisions and regional planning and finally SUP and FFH-VP integration.

A smooth and incremental transition to renewable electricity can be readily achieved by successively shutting down conventional power plants when they reach the end of their service lives and replacing these facilities with renewable electricity capacity. The transition scenarios presuppose that the mean service life of conventional power plants will be 35 years and that the current rate of renewable electricity expansion will be maintained. Both conditions can be regarded as rather restrictive. To do this the annual absolute capacity expansion rate for renewable electricity would have to increase to an average of 6 GW per year by 2020 (scenario 2.1.a). That rate is consistent with that achieved in past years. In the unlikely event that no electricity saving measures will be instituted, this figure would be 8 GW per year (scenario 2.1.b). The SRU estimates that even this elevated expansion rate could be successfully achieved by the relevant industries. But if German and EU energy efficiency objectives are steadfastly implemented or if the altogether realistic scenario comes true that operators of conventional fossil fuel fired power plants will want to keep their facilities in operation for longer than an average of 35 years, annual renewable electricity capacity expansion clearly below 6 GW will suffice between now and 2020.

These scenarios (2.1.a and 2.1.b) obviate the need to extend the service life of nuclear power plants or to build new coal fired power plants with CCS systems. In other words, the German existing fleet of conventional power plants, combined with a handful of newly built gas turbine plants, would provide a sufficient bridge for a transition to a wholly renewable electricity supply. Hence a transition scenario based on a relatively brief average service life amounting to 35 years for conventional power plants offers sufficient leeway and flexibility for the eventuality that the grid, storage capacity and generation capacity expansion for renewable energy will proceed more slowly than expected. A largely renewables-based system would have less of a need for base load power plants. The high volatility of renewables will necessitate a substantially higher level of flexibility on the part of all conventional power plants. The number of shutdowns required and rapid startup and shutdown procedures will rise in accordance with the increasing residual load and will obviate the need for a permanently available and consistent base load. Hence, once renewable electricity begins accounting for approximately 30 percent of aggregate electricity capacity, the construction of new conventional power plants will become unprofitable since it will no longer be possible to operate them at a sufficiently high capacity use level. Furthermore, if proportional renewable electricity use rises further still, base load power plant operation will become problematic from a technical standpoint as well. Moreover, extending nuclear power plant service life or building new coal fired power plants would entail the risk of surplus capacity over increasingly longer periods, thus necessitating renewable capacity downtime or cost-intensive underuse of conventional capacity and unnecessarily ramping up the costs of the transitional phase. Hence a blanket and pronounced extension of the service life of Germany's nuclear power plant fleet would be incompatible with the SRU scenarios involving a transition to a wholly renewable electricity supply.

Energy efficiency is a key precondition for transitioning to an affordable wholly renewable electricity supply. Certain legal and political measures would help Germany to transition to a wholly renewable electricity supply. Optimization and expansion of the emissions trading framework and of the Renewable Energy Act (EEG) are needed. In addition, the promotion of public participation in and acceptance of 100% renewable electricity supply are required. The regulatory aspects need to address how an optimized electricity transmission grid planning process can be achieved and which economic incentives for electricity storage facility and grid expansion are needed as support.[4]

[4] Further Reading: URL 1: Nr. 40, Materialien zur Umweltforschung, Möglichkeiten des großräumigen (transeuropäischen) Ausgleichs von Schwankungen großer Teile intermittierender Elektrizitätseinspeisungen aus regenerativen Energiequellen in Deutschland im Rahmen einer 100% regenerativen Stromversorgung mit dem Zeithorizont 2050, Dr. Gregor Czisch <www.umweltrat.de/SharedDocs/Downloads/DE/03_Materialien/2010_MAT40_Czisch.pdf>; URL 2: Nr. 41, Materialien zur Umweltforschung, Optionen der elektrischen Energieübertragung und des Netzausbaus, Technische Möglichkeiten und Kosten transeuropäischer Elektrizitätsnetze als Basis einer 100% erneuerbaren Stromversorgung in Deutschland mit dem Zeithorizont 2050, Prof. Dr. Heinrich Brakelmann, Prof. Dr. Istvan Erlich <www.umweltrat.de/SharedDocs/Downloads/DE/03_Materialien/2010_MAT41_Brakelmann_Erlich_Optionen_eletrische_Energie%C3%BCbertragung_Netzausbau.html>; URL 3: Nr. 42, Materialien zur Umweltforschung, Möglichkeiten und Grenzen der Integration verschiedener regenerativer Energiequellen zu einer 100% regenerativen Stromversorgung der Bundesrepublik Deutschland bis zum Jahr 2050 Deutsches Zentrum für Luft- und Raumfahrt <www.umweltrat.de/SharedDocs/Downloads/DE/03_Materialien/2010_MAT42_DZLR_Integration_Energiequellen_2050.html>; URL 4: Nr. 43, Materialien zur Umweltforschung, Planungs-, genehmigungs- und naturschutzrechtliche Fragen des Netzausbaus und der untertägigen Speichererrichtung zur Integration erneuerbarer Energien in die deutsche Stromversorgung, Professor Jens-Peter Schneider <www.umweltrat.de/SharedDocs/Downloads/DE/03_Materialien/2011_Mat43_Fragen_Netzausbau.html>; URL 5: Nr. 44, Materialien zur Umweltforschung, Systemkonflikt in der Transformation der Stromversorgung, Fraunhofer Institut für Windenergie und Energiesystemtechnik (IWES) <www.umweltrat.de/SharedDocs/Downloads/DE/03_Materialien/2010_MAT44_Systemkonflikt_Transformation_Stromversorgung.html>; URL 6: Sachverständigenrat für Umwelt (SRU) Aktuelle Stellungnahme Nr. 15: 100% erneuerbare Stromversorgung bis 2050: klimaverträglich, sicher, bezahlbar <www.umweltrat.de/SharedDocs/Downloads/DE/04_Stellungnahmen/2010_05_Stellung_15_erneuerbareStromversorgung.html>; Sachverständigenrat für Umwelt (SRU) Sondergutachten Wege zur 100% erneuerbaren Stromversorgung Erich Schmidt Verlag ISBN 978-3-503-13606-3, Pathways towards a 100% renewable electricity system <www.umweltrat.de/SharedDocs/Downloads/EN/02_Special_Reports/2011_10_Special_Report_Pathways_renewables.html> all accessed 17 July 2012.

Working Groups

International Climate Policy before COP 17

LEGAL OPTIONS FOR REGIME EVOLUTION IN THE CLIMATE CHANGE REGIME: SOME COMMENTS

Marc Pallemaerts[1]

1. Introduction

One of the most controversial issues faced by negotiators in the further development of the climate change regime is that of the legal nature and form of the new agreements to be concluded. As the negotiations are taking place within an established legal and institutional framework, they are constrained by the rules of that pre-existing framework, which determine the realm of legal possibility. The regime rules currently in force determine the options and procedures available for further regime development. These rules of the game must be taken seriously. As the Copenhagen experience shows, failure to do so entails systemic risks.

In this comment, we will be considering the range of options available and their implications from the perspective of international law and potential regime effectiveness. Our purpose here is to analyse the international legal framework for further regime development rather than to discuss the substance and the dynamics of the ongoing negotiations. By outlining the range of legal possibilities we merely intend to inform the debate on legal options and legal form which has become a central feature of the negotiations.

The climate regime in its current form comprises a framework convention (the UNFCCC) and an additional protocol (the Kyoto Protocol), which are legally independent instruments which do not have exactly the same contracting parties. Each of the two instruments is supplemented by hundreds of pages of detailed technical implementing rules laid down in the form of COP or CMP decisions. Though these rules are technically not legally binding, many of them—especially those based on a specific enabling provision—carry a strong expectation of compliance.

The UNFCCC provides two different procedures for the further development of the climate change regime: new provisions can be adopted

---

[1] Professor of European Environmental Law, University of Amsterdam.

either in the form of amendments to the Convention itself, or in the form of additional Protocols. And, where a Protocol is in force, that instrument itself can also be amended in accordance with the rules it provides. This brings us to the first option for further regime development which is currently on the negotiating table: amending the Kyoto Protocol.

2. Options for Regime Development

2.1. *Amendment of the Kyoto Protocol*

The history of the Kyoto Protocol goes back to the Berlin Mandate in 1995. Following the first scheduled review of the adequacy of the mitigation commitments of Annex I Parties to the UNFCCC in 1995, COP3 in 1997 eventually decided to take further action by adopting an additional Protocol, which became the Kyoto Protocol, rather than by amending the provisions on commitments in art. 4(2) of the Convention. Both options were originally on the table, but the Protocol option eventually prevailed, resulting in the dual-track regime design that we currently have.

The Kyoto Protocol lays down additional, quantified mitigation commitments for Annex I Parties which have ratified it for a period starting in 2008 and ending in 2012. These so-called Annex B Parties to the Protocol will have no further or continued quantified commitments under the Protocol beyond 2012 unless the Protocol itself is amended, preferably before the end of its first commitment period.

A procedure for the adoption of further commitments is provided in art. 3(9) of the Protocol, which stipulates that commitments for subsequent periods for Parties included in Annex I shall be established in amendments to Annex B, the annex which sets forth the QELRCs of individual Parties. COP/MOP1 decided to initiate the consideration of such commitments in 2005, but, more than six years later, these negotiations have not yet been concluded, and it has become certain that the Protocol cannot be amended in time to avoid a legal gap between the first and second commitment periods.

Should agreement be reached on post-2012 commitments, Annex B of the Protocol shall be amended in accordance with art. 21(7) of the Protocol, which requires the written consent of each individual Party with a commitment in Annex B, followed by the deposit of instruments of acceptance by the Parties. The amendment thus adopted will only enter into force for those Parties which have individually and formally accepted it, provided at least three fourths of the Parties to the Protocol confirm their acceptance of the amendment. It is important to recognise that even the formal adoption of an amendment to Annex B does not guarantee that all initial Annex B Parties will also be bound by commitments post-2012.

The drafters of the Kyoto Protocol assumed that future commitments would be established in the same form and within the same framework as the initial commitments for the first commitment period, i.e. merely by amending Annex B, although the Protocol does not specify that subsequent periods should necessarily have the same five-year duration as the first one. However, since the procedural requirements for amending Annex B are effectively the same as for any other amendments to the provisions of the Protocol, there is nothing to prevent Parties from agreeing on changes in the basic framework and other provisions of the Protocol as part of the same round of negotiations which was launched in 2005. However, this is an uphill struggle, as there has been a strong tendency on the part of non Annex I Parties to the Kyoto Protocol to resist the opening of negotiations on any aspect of the Protocol other than the provisions in Annex B and minor consequential amendments. These Parties fear that opening wider negotiations may lead to a situation in which they might also be called upon to subscribe to mitigation commitments.

This is precisely why choosing an amendment of the Kyoto Protocol as the main avenue for further development of the climate regime beyond 2012 carries important legal and political constraints, in view of the opposition to fundamental change in the structure of the Protocol by non Annex I Parties, the fact that the negotiation excludes non-Kyoto Parties and the fact that any amendments can only apply to Protocol Parties and not to all UNFCCC Parties. That is why Annex I Parties to the UNFCCC, in Bali, successfully pressed for the opening of a second set of negotiations, in parallel with the Kyoto amendment talks. These parallel negotiations conducted under the UNFCCC, are much broader in scope and their potential outcome not as rigidly defined. Under the mandate laid down in the Bali Action Plan reference is made only to an 'agreed outcome' without further details as to its legal form or nature. As the negotiations are conducted within the framework of the UNFCCC, again, there are potentially two forms in which a legally binding outcome could be captured: a new Protocol or an amendment to the Convention itself. In the remainder of this comment, we will consider both options in turn.

2.2. *New Protocol under the UNFCCC*

According to Art. 17, a Protocol to the Convention can be adopted at any ordinary session of the COP, in accordance with its general rules of procedure, which imply that the decision to adopt a Protocol shall be taken by consensus. The requirements for the entry into force of any such Protocol are to be established by the instrument itself. Thus, these requirements may differ from those laid down in 1997 for the entry into force of the Kyoto Protocol as well as from those applicable to amendments to the Convention. This procedure for regime development therefore provides maximum flexibility, including for the possible development, within the overall framework of the UNFCCC, of agreements for cooperative action that would not necessarily involve all the Parties. Indeed, Art. 17 could be used in conjunction with Art. 7(2)(c), which mandates the COP to 'facilitate, at the request of two or more parties, the coordination of measures adopted by them to address climate change and its effects'. It would be possible for the COP to adopt more than one new Protocol simultaneously, which would provide Parties with a choice in terms of ratification. Since the opening of the negotiations under the Bali Action Plan, several Parties have formally tabled proposals for new Protocols under Art. 17, but none of these proposals has effectively become the focus of negotiations so far. In Cancun, COP16 called for due attention to be given to these proposals along with the other negotiating texts on the table.

Many parties consider that ultimately legally binding outcomes will also have to be reached in the UNFCCC track in order to ensure not only balance with the Kyoto Protocol track but also effectiveness of mitigation and adaptation action worldwide. Thus, the question of the legal form of the outcome has become increasingly salient and negotiators have started considering options for legally binding outcomes. In this debate, new Protocols seem to be the preferred option, but the preference of many Parties for this approach should not lead negotiators to disregard the other option available to them: amending the UNFCCC.

2.3. *Amendment of the UNFCCC*

The choice of the alternative procedure of an amendment to the Convention itself may well be worth more serious consideration than it has received so far, compared to the Protocol option. It has both advantages and disadvantages.

The amendment procedure might well make it easier to craft a comprehensive package deal that would be acceptable to the widest possible range of Parties, without giving any individual Party a veto, since unanimity is not required for the adoption of amendments (although no Party can be bound by an amendment without its consent). Another advantage is that it would bring all Parties together under a single legal framework, rather than a patchwork of protocols. However, this advantage would not easily materialise because it would require ratification of the amendment by all Parties. However, as the experience of the Montreal Protocol shows, the fact that some Parties may not be bound by a particular amendment should not necessarily lead to serious difficulties in regime implementation.

One should recall that the very notion of regular reviews followed by possible amendments was written into the UNFCCC from the start. Art. 4(2)(d) of the Convention provides for reviews of the commitments of Annex I Parties at regular intervals, possibly leading to their amendment. It is on the basis of this provision that COP1 decided to initiate the negotiation process which eventually led, not to an amendment of the Convention, but to the adoption of the Kyoto Protocol by COP3 in 1997. Though the same provision required a second review by the COP to take place not later than 31 December 1998, none has been carried out due to lack of consensus on the scope of this exercise and its link with the issue of future commitments. Art. 7(2)(a) also gives the COP a general mandate to periodically examine the obligations of the Parties and the institutional arrangements under the Convention, in the light of the objective of the Convention, the experience gained in its implementation and the evolution of scientific and technological knowledge.

The lack of success in implementing these review provisions is directly linked with the main disadvantage of the amendment option: the fear on the part of non Annex I Parties that any amendment of the UNFCCC would lead to a renegotiation of the 'firewall' between developed and developing countries enshrined in the current provisions. This lack of confidence in the regime makes many Parties reluctant to contemplate the option of an amendment which could, in their view, lead to a 'reopening' of fundamental Convention provisions. Incremental regime change through new Protocols is perceived as less risky than amendment of the basic instrument itself. For this reason, the amendment option, despite its advantages from a legal and procedural viewpoint, may well have to be ruled out on political grounds.

INTERNATIONAL CLIMATE CHANGE POLICY: AN INDIAN PERSPECTIVE

Lavanya Rajamani[1]

1. Introduction

India is on a mission to develop. Economic growth, and with it, poverty eradication, energy security, and provision of universal access to energy, are central and enduring preoccupations of the Indian government. Justifiably so: India is placed 134th on the Human Development Index,[2] 41.6 per cent of its population lives on less than 1.25 US$ a day,[3] and an estimated 44 per cent does not have access to electricity.[4] India's energy use is currently at a low per capita emissions rate of 1.5 metric tons annually,[5] and a cumulative share of 4.6 per cent.[6] If India's current growth rate—8%—continues, however, energy demand will more than double by 2020, and India will soon be a significant contributor to climate change. India, and its development agenda, will also, it is believed, be severely affected by climate change impacts. These considerations frame the tenor, tone and substance of India's engagement with the international climate regime.

India is a Party to the Framework Convention on Climate Change (FCCC), 1992, and its Kyoto Protocol, 1997. Indian negotiators were instrumental in crafting the principle of 'common but differentiated responsibility', a particular interpretation of which provided the blueprint for the controversial differential treatment contained in the Kyoto Protocol. India has insisted, in coalition with other developing countries, that the Kyoto Protocol continues past the end of its first commitment period in 2012 i.e.

---

[1] Professor (International Law), Centre for Policy Research, New Delhi.

[2] UNDP, 'Human Development Report: Overcoming Barriers' (2009) <http://hdr.undp.org/en/statistics> accessed 03 July 2012.

[3] Ibid.

[4] UNDP, 'Human Development Report: Fighting Climate Change' (2007) <http://hdr.undp.org/en/reports> accessed 03 July 2012.

[5] Indian Network for Climate Change Assessment, Ministry of Environment and Forests, Government of India, 'India: Greenhouse Gas Emissions 2007' (May 2010). All Ministry of Environment and Forests documents on climate change are available at <http://moef.nic.in/modules/about-the-ministry/CCD> accessed 03 July 2012.

[6] UNDP (n. 3).

that developed countries taken on legally binding GHG mitigation targets for a subsequent commitment period. India is also insistent that no legally binding quantitative GHG mitigation targets attach to itself given India's limited role in contributing to the problem, its overriding development needs, and the historical responsibility of developed countries.

2. India's Negotiating Position

In the negotiations launched by the Bali Action Plan, 2007, under the auspices of the Ad Hoc Working Group on Long-term Cooperative Action (AWG-LCA), India has opposed a quantitative long-term global goal or a peaking year, unless accompanied by an appropriate burden sharing arrangement based on equity and differential treatment for developing countries. In Cancun, India as part of the BASIC (Brazil, South Africa, India and China) countries introduced the term 'equitable access to sustainable development' in this context.[7] India has since spear-headed efforts within the BASIC group to flesh out and operationalize this notion. A BASIC Experts Paper on Equity was presented at the 17th Conference of Parties to the FCCC at Durban, December 2011.[8] India perceives 'equitable access to sustainable development' as signaling the recognition that developing countries need to grow, albeit sustainably, and that there must be an equitable allocation of the global carbon budget to enable them to do so.[9]

Nevertheless, India has offered to subject its national actions to more frequent reporting as well as to a process of 'international consultation and analysis'. India also offered under the non-binding Copenhagen Accord[10] to 'endeavour to reduce the emissions intensity of its GDP by 20–25% by 2020 in comparison to the 2005 level'.[11] This undertaking has

[7] Decision 1/CP.16, 'The Cancun Agreements: Outcome of the work of the Ad Hoc Working Group on Long-term Cooperative Action under the Convention', FCCC/CP/2010/7/Add.1 (15 March 2011), para 6.

[8] BASIC Experts, 'Equitable access to sustainable development, Contribution to the body of scientific knowledge' (2011) <www.erc.uct.ac.za/Basic_Experts_Paper.pdf> accessed 03 July 2012.

[9] Proposals by India for inclusion of additional agenda items in the provisional agenda of the seventeenth session of the Conference of the Parties, Note by the Secretariat, FCCC/CP/2011/INF.2/Add.1 (7 October 2011), Annex.

[10] Decision 2/CP.15, Copenhagen Accord, FCCC/CP/2009/11/Add.1 (30 March 2010), 4.

[11] India—Letter to the Executive Secretary (30 January 2010) <http://unfccc.int/files/meetings/cop_15/copenhagen_accord/application/pdf/indiacphaccord_app2.pdf> accessed 10 October 2012.

been mainstreamed into the FCCC process through an information document taken note of[12] by the Cancun Agreements, 2010.[13]

3. India at Durban

The debate on equity however has proven intractable and India had been hesitant until Durban to agree to negotiate a new legally binding instrument. India, with China, had argued that 'substance must precede form,' and that until there is greater clarity on the content of the obligations that are necessary, a discussion on legal form is premature. At Durban, however, after protracted negotiations, India agreed to participate in a decision to launch a process titled the *Durban Platform on Enhanced Action* to negotiate 'a Protocol, another legal instrument or agreed outcome with legal force under the Convention applicable to all.'[14] This instrument is intended to be adopted in 2015, and implemented from 2020. Although the media and some developed countries have interpreted this decision to mean that all 'major emitters,' developed and developing alike, will be legally bound to GHG emissions cuts from 2020,[15] the language used in this decision masks a vast terrain of contestations.

India, until the final hours of the conference, had insisted that agreeing to a legally binding instrument was a red line that it could not cross. It could agree at best to launch a process towards a 'legal outcome'—which would leave the precise legal form of the instrument open. A 'legal outcome' could encompass legally binding instruments as well as Conference of Parties decisions, which although operationally significant, are not, save in the exception, legally binding. This formulation lacked the clarity and ambition that the EU, the Alliance of Small Island States, the Least Developed Countries, many Latin American countries, and even India's BASIC allies, Brazil and South Africa, were seeking. Critically, this

---

[12] See Compilation of information on nationally appropriate mitigation actions to be implemented by Parties not included in Annex I to the Convention, FCCC/AWGLCA/2011/INF.1 (18 March 2011), 26.

[13] Decision 1/CP.16, 'The Cancun Agreements: Outcome of the work of the Ad Hoc Working Group on Long-term Cooperative Action under the Convention', FCCC/CP/2010/7/Add.1 (15 March 2011), para. 49.

[14] Decision 1/CP.17, 'Establishment of an Ad Hoc Working Group on a Durban Platform for Enhanced Action', FCCC/CP/2011/9/Add.1 (15 Mar. 2012).

[15] Todd Stern, 'Durban: An Important Step Forward in Combating Global Climate Change' (*Huffington Post*, 16 December 2011) <www.huffingtonpost.com/todd-d-stern/durban-climate-talks_b_1153721.html> accessed 03 July 2012.

was not sufficient for the EU to endorse a second commitment period of the Kyoto Protocol. After a fast and furious 'huddle' in the final hours of the conference, India agreed to substitute the term 'legal outcome' with a marginally less ambiguous term, 'agreed outcome with legal force,' thus triggering the acceptance of a Kyoto second commitment period by the EU and its allies.[16]

It is significant, given India's insistence on equitable burden sharing, that the Durban Platform decision does not contain a reference to equity or common but differentiated responsibility. Through the two weeks of the conference, developed countries were unanimous in their insistence that any reference to 'common but differentiated responsibilities' must be qualified with a statement that this principle must be interpreted in the light of 'contemporary economic realities.' They were also insistent that the future regime must be 'applicable to all.'[17] India, among other developing countries, argued in response that this would be tantamount to amending the FCCC. The text was eventually drafted such that it was rooted in the Convention—'under the Convention'[18]—thereby implicitly engaging its principles, including the principle of common but differentiated responsibilities. This, India believes, would hold efforts to reinterpret and qualify this principle at bay.[19]

4. India's Domestic Policies & Actions

Notwithstanding its seemingly 'obdurate' stance in the international negotiations,[20] India has introduced a raft of measures domestically in the last few years. In 2008, India launched its National Climate Change Action Plan bringing together existing and proposed efforts at decarbonisation under eight national missions: solar energy; enhanced energy efficiency; sustainable habitats; water; the Himalayan ecosystem; sustainable agriculture; and strategic knowledge for climate change.[21] The relevant Ministries

[16] Lavanya Rajamani, 'Deconstructing Durban' (*Indian Express*, 15 December 2011) <www.indianexpress.com/news/deconstructing-durban/887892/> accessed 03 July 2012.

[17] n. 13, para 2.

[18] Ibid.

[19] n. 15.

[20] 'China, India and climate change: Melting Asia' *The Economist* (Beijing, Delhi, Tibet, 5 June 2008) <www.economist.com/node/11488548> accessed 03 July 2012.

[21] Prime Minister's Council on Climate Change, Government of India, 'National Action Plan on Climate Change' (2008), <www.indiawaterportal.org/node/9781> accessed 03 July 2012.

have since developed comprehensive mission documents detailing objectives, strategies, plans of action, timelines, and monitoring and evaluation criteria.[22] There are several noteworthy initiatives contained in these missions, including: the creation of a market—a perform, achieve and trade mechanism—in energy savings certificates; the adoption of a target to generate 20,000 MW of solar power by 2022; and a commitment to double the area to be afforested in the next 10 years, taking the total to 20 million ha.[23] In addition, the Indian Government has announced a levy—a clean energy tax—of US$1 per ton on coal.[24] State-level action plans on climate change are also in preparation.[25]

India has through these, and other, measures demonstrated a commitment to addressing climate change and its impacts. India has also demonstrated a willingness to provide transparency to its international partners in relation to these actions. At Durban, India further agreed to negotiate a new 'agreed outcome with legal force' for the post-2020 period. Anything further from India, in particular GHG mitigation targets in such a post-2020 legally binding instrument, however will depend on the willingness of the international community to arrive at an equitable sharing arrangement that reassures India that it will not be mortgaging its future to address a problem not, at least in origin, of its making.

---

22 Ministry of Environment and Forests, 'Finalisation of the Eight National Missions' (Press Information Bureau Release, 11 August 2010) <http://pib.nic.in/newsite/erelease.aspx?relid=64607> accessed 03 July 2012.

23 n. 10.

24 Ibid., 2.

25 State level action plans of 10 states have been provisionally approved by the Ministry of Environment and Forests. These are available at <http://moef.nic.in/modules/others/?f=sapcc-2012> accessed 03 July 2012.

INTERNATIONAL CLIMATE LAW AND POLICY: AN AUSTRALIAN PERSPECTIVE

Jacqueline Peel[1]

Climate change is an issue of great environmental and political salience for Australia. As an already hot and arid continent, the impacts of climate change for Australia are predicted to be relatively severe compared with many other developed countries.[2] Addressing climate change in law and policy also poses major socio-economic challenges for Australia. Australia's economy is highly energy-intensive and reliant upon a stationary power sector in which over 90 per cent of electricity is generated from burning fossil fuels, principally brown and black coal.[3] Australia's per capita emissions are the highest in the OECD and among the highest in the world as a consequence of its emissions-intensive energy production and transportation processes.[4] Australia is also heavily dependent upon exports of coal and gas for national income: the country is the world's second largest exporter of thermal coal, and coal is among Australia's top exports, accounting for 15 per cent of Australia's total exports by value.[5]

Against this backdrop, the politics of climate change regulation in Australia have often been fraught. Internationally this saw Australia pursue, and secure, during the Kyoto Protocol negotiations, special concessions including a generous emissions reduction target of an 8% *increase* on 1990 levels by 2012,[6] and provision to allow the inclusion of emissions from land-clearing in calculation of the nation's 1990 baseline emissions.[7] In general, the framing of Australian climate change policy has been heavily

---

[1] Professor Jacqueline Peel, Melbourne Law School, University of Melbourne, Australia.

[2] See CSIRO, Climate Change in Australia: Technical Report 2007: Executive Summary (2007); CSIRO, Fact Sheet: Australia is vulnerable to climate change, 2 <http://www.csiro.au/resources/Climate-Change-Vulnerable.html> accessed 04 July 2012.

[3] Rodney Tiffen and Ross Gittens, *How Australia Compares* (2nd ed, Cambridge University Press 2009) 158.

[4] Ross Garnaut, *The Garnaut Climate Change Review* (Cambridge University Press 2008) 153.

[5] Department of Foreign Affairs and Trade, Composition of Australian Trade 2011 (2012) 4 and 12.

[6] Kyoto Protocol, Annex B.

[7] Kyoto Protocol, Article 3.7. Tellingly, this provision is known as the 'Australia clause'.

dependent on developments (or the lack thereof) taking place at the international level and close attention has been paid to what other countries and regions—particularly more powerful economies such as the European Union and the United States—are doing to reduce emissions of greenhouse gases.

From an Australian perspective, the first best approach to climate change regulation would be through conclusion of a comprehensive international agreement, that incorporates all major emitters (including developing countries such as India, China and Indonesia), and seeks to stabilise emissions at levels sufficient to ensure global warming reaches no more than 2°C above pre-industrial levels. This goal has been consistently reiterated in Australian foreign policy statements and in the country's contributions to the international climate change negotiations.[8] However, with the failure of the Copenhagen conference in 2009 to agree on emissions reductions post-2012 and the tepid progress made at the international level since then, this objective looks increasingly more difficult to realise. In policy and academic circles in Australia—as elsewhere around the globe—attention is turning to how climate change regulation might work in a 'bottom-up' fashion, with contributions at the national and regional levels driving legal and policy reform, rather than 'top-down' international legal obligations.

While there are strong political and economic considerations that might dictate a cautious approach to climate change regulation in Australia in the absence of collective global action to reduce greenhouse gas emissions, Australia has recently moved to introduce a wide-ranging policy and legislative package for achieving climate change mitigation. The package—unveiled by the Australian Prime Minister in July 2011, with the first suite of implementing legislation passed by the federal parliament in November 2011—includes for the first time in Australia a national carbon pricing mechanism that applies to a range of major carbon polluters in the stationary energy, waste, rail, domestic aviation and shipping, mining and industrial processes sectors from 1 July 2012.[9] The carbon pricing mechanism

[8] See Australia, Legal Architecture for a Post-2012 Outcome: Submission to the AWG-LCA and AWG-KP, available at <www.climatechange.gov.au/government/international/~/media/submissions/international/Legal-Architecture-Post-2012-Outcome-Australian-submission-MAY-09.pdf> accessed 04 July 2012. See also Australia-Norway, Submission under the Cancun Agreements: Enhanced Action on Mitigation, AWG-LCA/AWG-KP, September 2011, at <http://unfccc.int/files/meetings/ad_hoc_working_groups/lca/application/pdf/australia_norway_mitigation_submission_.pdf> accessed 04 July 2012.

[9] Clean Energy Act 2011 (Cth).

initially has a fixed price period of three years, transitioning to a fully flexible emissions trading scheme (ETS) from July 2015 onwards. While the initial emissions reduction target committed to by the Australian government is extremely modest—5% below 2000 levels by 2020—there is capacity for the emissions cap to be adjusted during the fully flexible period of the scheme in light of international developments, including more ambitious action on climate change mitigation at the global level or in other countries and regions. In this respect, an important element of the package is the establishment of a Climate Change Authority—modelled on the UK Committee on Climate Change—to advise the government on appropriate settings for the scheme's emissions caps.[10]

Reflecting a more general trend in climate law and policy, the Australian government's policy package goes significantly beyond an ETS. Other elements of the package include significant measures for investment in renewable energy, new requirements with respect to energy efficiency and measures to encourage reduction of emissions in the land-use and agricultural sectors (another major contributor to greenhouse gas emissions in Australia).[11] These measures will sit alongside existing climate laws, including a target for renewable energy generation applicable to entities in the electricity generation sector,[12] and other regulatory initiatives undertaken by State governments. While the policy package is far from perfect (e.g. the exclusion of most forms of transportation from coverage under the carbon pricing mechanism at least until 2014 and the extensive provision made for compensation to trade-exposed, emissions-intensive industries and the coal-fired power generation sector) at this early stage it appears that Australia has taken an important step towards making the social, legal and economic changes that will be necessary for the country to reduce its high levels of greenhouse gas emissions.

The political road that led Australia to this point has been tortuous and the role of the Australian Greens party (who have the balance of power in the Upper House of the Australian Parliament and lend support to the Labor government in the Lower House) should not be underestimated as a factor pushing the national government to adopt its current climate

---

10 Climate Change Authority Act 2011 (Cth).

11 Australian Government, Securing a Clean Energy Future: The Australian Government's plan (2011). The mechanism for generating carbon credits from agricultural activities—known as the Carbon Farming Initiative—was legislated by the Carbon Credits (Carbon Farming Initiative) Act 2011 (Cth).

12 Renewable Energy (Electricity) Act 2000 (Cth).

change mitigation measures. Ironically the consequence of release of this package may be to transform Australia from a regulatory laggard in climate change terms to a pioneer, albeit a reluctant one. In the increasingly fragmented, 'bottom-up' regime of global climate change law and policy that has emerged post-Copenhagen this approach poses both risks and potential benefits for Australia. If other countries do not follow suit in adopting climate change mitigation measures, then Australia's actions may come at a high national economic cost with little environmental benefit given the country's low emissions (as a percentage of global emissions). On the other hand, if Australia's actions provide momentum to those of other countries and to international negotiations, Australia stands to reap enormous benefits as a 'first mover' on climate change regulation, both in terms of making the transition to a low carbon economy and in ensuring that the globe avoids dangerous climate change that will have particularly severe consequences for Australia.

DISCUSSION SUMMARY
WORKING GROUP: INTERNATIONAL CLIMATE POLICY BEFORE COP 17

Sebastian Oberthür[1]

Discussions in the working group were led by a panel composed of five distinguished speakers: Lavanya Rajamani of the Centre for Policy Research (India), Daniel Bodansky of Arizona State University (USA), Jacqueline Peel of Melbourne Law School (Australia), Marc Pallemaerts of the University of Amsterdam (the Netherlands), and Dirk Messner of the German Development Institute (Germany). On the eve of the 17th Conference of the Parties to the United Nations Framework Convention on Climate Change (UNFCCC) and the 7th Conference of the Parties serving as the meeting of the Parties to the Kyoto Protocol, discussions focused on two core themes: (1) the future of the Kyoto Protocol and (2) the respective merits of top-down versus bottom-up approaches. They proceeded against the backdrop of a widening gap between what science tells us would be required to prevent 'dangerous anthropogenic interference with the climate system' in accordance with Article 2 of the UNFCCC and what is happening in reality.

After the preceding keynote speeches by Bodansky and Messner, the first two panel contributions by Rajamani and Peel focused on national-level action in and positions of India and Australia, respectively. Rajamani started by referring to the overall policy context which forms the background for Indian climate politics. Thus, India is placed 134th on the Human Development Index and 44% of its population, i.e. 500 million people, do not have access to electricity. India's current annual growth rate is 8–9%; in order to sustain 8% annual growth through 2031, India would need to increase its energy supply 3–4-fold. At the same time, per capita emissions in 2007 were still relatively low at 1.5 tonnes and India is facing serious impacts of climate change; in the words of the former environment minister of India Ramesh: 'no country in the world is as vulnerable as India'. Rajamani then presented the key requests of India in

---

[1] Prof. Dr. Sebastian Oberthür, Vrije Universiteit Brussel, Institute for European Studies, Belgium.

the international negotiations, including respect for the principle of 'common but differentiated responsibilities', retaining the 'firewall' between developed and developing countries and continuing the Kyoto Protocol. According to India, action by developing countries shall differ from those of developed countries in terms of their *legal form* as well as their *character* and *stringency*. India would agree to a long-term quantitative global goal or peaking year, but only if there was agreement on equitable access to sustainable development (i.e. on an international burden sharing basis). No country should enact unilateral trade measures against other countries on climate grounds. Regarding a new legally binding instrument, India is rather resistant, emphasising that substance should come before form. Finally, Rajamani stressed that India had made important strides internationally as well as domestically to address climate change. She pointed out that India had agreed to a greater level of transparency. While India had so far only submitted one 'National Communication' to the UNFCCC, it will from now on, as stipulated in the Cancun Agreements, submit a National Communication every four years with additional biennial updates. Agreeing to 'international consultation and analysis' of domestic climate action was also a great step for India. In terms of domestic climate policy, Rajamani held that a number of measures including eight so-called 'national missions' (among others, for solar energy, energy efficiency, and forests) and a recently announced carbon tax of 1 US$ per tonne of coal demonstrated that India is truly committed to the cause of climate protection. Overall, Rajamani argued that India has become an important deal-maker in the international arena.

Peel subsequently presented her Australian perspective and in particular cautiously hypothesised that Australia might evolve from a climate laggard to a reluctant pioneer. She presented proposed legislation before the Australian Parliament (and adopted later in 2012, subsequent to the discussion in the working group). It especially foresees the establishment of a national carbon trading scheme. For the first three years, the carbon price would be fixed (at 23 Australian Dollars) and thus equal a carbon tax. Subsequently, the system would be up-scaled into a full emissions trading scheme. The current Australian target is for a 5% reduction of greenhouse gas emissions below 2000 levels by 2020, with the possibility of a strengthening of this target. Peel highlighted the massive public debate surrounding climate policy, including public demonstrations against a carbon tax. She furthermore pointed to Australia being a country that is both very vulnerable and contributing to climate change. On the one hand, Australia has recently experienced several extreme weather incidents such as heavy rainfalls, droughts, flooding, and cyclones. On the other hand, Australia is heavily dependent on burning fossil fuels (90% of its electricity comes from fossil fuels) and an important exporter of coal. With regard to international climate policy, Australia is set to achieve its modest emission target of +8% above 1990 under the Kyoto Protocol. Compared with this target, the proposed legislation constitutes a significant step forward. More ambitious domestic climate policies to a large degree depend on international progress. Peel concluded that through its domestic policy Australia might regain moral authority and that progress achieved domestically might also have a signalling effect for the international negotiations.

In his subsequent intervention, Pallemaerts, responding in particular to the keynote presentation given by Bodansky in the plenary session, focused on the international legal framework to combat climate change. He first took issue with Bodansky's call for instrumental diversity and argued that the existing legal system already provides for instrumental diversity, with ratified commitments presented in a targets-and-timetable approach under the Kyoto Protocol, less explicit commitments in the UNFCCC and a body of COP decisions that are technically but not legally binding. He further argued that the stark dichotomy between top-down and bottom-up approaches is a caricature since there is no real top-down decision-making in international law. The quantified emission limitation and reduction commitments of developed countries under the Kyoto Protocol were not the result of a top-down process, but rather of a bottom-up process, in which pledges by different parties were couched into treaty language. A top-down process would even be unfeasible due to the consensus rule that governs decision-making in the international climate regime. Furthermore, Pallemaerts argued against attempts to relativise the relevance of international law. He presented the EU as committed to 'effective multilateralism' under the Lisbon Treaty (i.e. the Treaty on European Union and the Treaty on the Functioning of the European Union) citing in particular the EU Treaty's Article 21(1). Turning to possible ways forward for the international cooperation on climate change, Pallemaerts argued that there would not be any progress if Parties abandoned the Kyoto Protocol. Parties to the UNFCCC might pursue several options, including (1) amending the Kyoto Protocol, (2) amending the UNFCCC, (3) adopting a new protocol, and (4) adopting COP decisions. Since no single party can veto amendments to the UNFCCC, this option might be feasible. Finally, Pallemaerts added that the EU should engage in an equity debate about the sharing of carbon space, since this is 'the crux of the issue'.

Complementing his keynote presentation in the plenary session, Messner offered three additional thoughts. First, while conceding that more flexible / polycentric approaches can contribute to a solution, they should only be a supplement to the existing framework. Focusing on what countries are currently willing to do will not satisfy the needs as presented by science. While it is important that we move in the right direction, this alone will be insufficient in the case of climate change. Given the small carbon budget still available and the time constraint, moving into the right direction alone would in all likelihood mean that the limit of a maximum of an average warming of 2 degrees Celsius, as established by science and policy-making, will be exceeded. Second, he argued that countries and societies cannot avoid the burden-sharing debate. Climate politics involves a distributional conflict. Trying to avoid addressing this conflict now will only mean that it will be shifted into the future. Third, he emphasized the importance of market signals towards a great transformation. Investments in a low-carbon economy will be too expensive if there are no appropriate market signals and incentives. Investments in innovation need clear signals for business players. Such signals can include carbon taxes, carbon trading, subsidies and others.

Responding to Pallemaerts, Bodansky countered the notion that the US is the only country acting unilaterally. He argued that the EU, in his view, also acts unilaterally, e.g. through the inclusion of aviation in the EU emissions trading scheme (although it was pointed out that the EU only moved ahead after having tried to achieve a multilateral solution in the International Civil Aviation Organisation (ICAO) for 10 years). He further clarified that he did not see a political agreement (as opposed to a legal instrument) as the most desirable outcome, but as the most realistic option given that 'we are not in a perfect world'. As regards bottom-up versus top-down approaches, he held that the UNFCCC and the Kyoto Protocol constitute two very different approaches: Whereas the Kyoto Protocol contains very specific targets and timetables which result from a rather top-down way of thinking, the UNFCCC is more open towards all different sorts of domestic policies and actions. Furthermore, he argued that the emission targets of the Kyoto Protocol were not the result of mere pledges. For example, the USA went into Kyoto with a different bottom-line position, but then agreed to a more ambitious target in the course of the negotiations. Overall, if a top-down approach can only capture less than 25% of world's emissions, then everyone should think about other approaches. In his view, there is a choice between trying more realistic options and sticking with something that does not work in practice.

The ensuing open discussion within the working group pushed the debate among the panellists further. Contributing to the debate on the prospects of bottom-up developments, it was clarified that the Australian government intends to link their emerging emissions trading scheme with other relevant systems, possibly including the EU emissions trading scheme. However, official bilateral negotiations between the EU and Australia were not going on at the time of the debate, and participants acknowledged challenges inherent in attempts to link different trading systems such as comparability of targets / caps. Peel furthermore highlighted in the debate that pursuing a bottom-up approach does not necessarily constitute an easy option, since the Australian Prime Minister was risking her political life by pushing for a domestic emissions trading system.

Another crosscutting theme of the discussion concerned the tension between what is scientifically required and what is politically feasible. How can an ambitious solution to climate change be found, while also recognising the constraints resulting from the political reality? Since there is no easy answer available to this question under the current conditions of (international) politics of climate change, it was not surprising that nobody had the panacea for addressing this challenge and dilemma. It was pointed out, however, that an international academic conference should put forward solutions that are clearly in line with science.

In this context, Messner suggested that the drive for ambitious climate policies will not come from the international negotiations but from economic measures and incentives initiated by key countries. He pointed to a number of encouraging signs and developments highlighting that both within industries and societies important shifts into the right direction are already taking place. For example, in Germany there are now more people employed in the 'green' sector of the economy than in the car industry. As to the international negotiations, he suggested assembling the first 30 movers together in a WTO style grouping might be an option worth thinking about.

Several other participants highlighted the need and rationale for continuing with the Kyoto Protocol. It was pointed out that abandoning the Kyoto Protocol would mean to lose not only the treaty text itself, but also the detailed rules contained in the numerous subsequent decisions, especially the Marrakesh Accords of 2001. Calls for flexibility should not serve as an excuse for abandoning the Kyoto Protocol, since the latter provides for a high level of flexibility. It was pointed out that 'we do not have time to reinvent the wheel'. The continuation of the Kyoto Protocol would also put the continuation of its market mechanisms, most importantly the Clean Development Mechanism, beyond doubt (even though future market demand for CDM credits remains very much uncertain).

On the form of future international action on climate change, a consensus seemed to be emerging that, *ceteris paribus*, a legally binding approach would be most effective. However, several participants remained sceptical as to the political feasibility of a new international legal instrument covering all major emitters. In addition, Bodansky pointed to the importance of other variables, beyond the legal nature of an agreement, including its precision, stringency, etc. As to the political feasibility of an international agreement, scepticism was also expressed as regards the role of the United States and its ability to actively contribute to and participate in any international framework (even if consisting of several 'docking stations' to be used flexibly) and to show international leadership. Following this line of thought, the world would have to move ahead regardless of what the United States does. Finally, several speakers concurred with Pallemaerts and Messner in highlighting the need to address the issue of equity in more detail in the international process.

EU and National Initiatives

CLIMATE PROTECTION: A SOUTH AMERICAN PERSPECTIVE

Jorge Caillaux[1]

1. Our Vision Concerning the International Process on Climate Change

As a member of the Latin American Platform on Climate[2] we believe there is neither justification nor time for developing countries to remain inactive, as we have a certain degree of responsibility regarding this problem and must be prepared to increase our efforts towards mitigation of greenhouse gas emissions.

Developing countries understand that the problem of climate change[3] does not originate from current emissions, but from the historic accumulation of greenhouse gases. They demand that developed countries undertake effective efforts to reduce emissions in accordance with their responsibility. As Gerardo Honty says:

> In brief and very simply: rich countries will not commit money nor agree to greater compromises until they see reduction commitments by poor countries as well. On the other hand, developing countries will not reduce their own emissions until they receive financial support and the industrialized countries show clear signs of significantly addressing their emission output.

---

[1] Dr. Jorge Caillaux, Peruvian Society for Environmental Law (SPDA).

[2] The 17 member organizations of the Latin American Platform on Climate (LAPC), from nine countries in the region, created in January 2009, identified a gap in Latin Americas intellectual production and reflection regarding climate change, capable of generating responses in the context of the global climate change phenomenon. In order to overcome this gap, as a key objective of its actions, the Platform decided to focus its attention on climate change and its effects, and give priority to policy, environmental, social and economic decision making at multiple levels, both in the public and private sectors of the region <www.intercambioclimatico.com> accessed 06 July 2012.

[3] The present statement is based in several publications of Dr. Gerardo Honty, principal researcher of CLAES and a special consultant of the Latin American Platform on Climate. For a complete analysis see: Gerardo Honty, *Cambio Climático: Negociaciones y consecuencias para América Latina* (2010).

2. The Case of Latin America

Unlike what is occurring in the rest of the world, the greatest proportion of emissions in Latin America do not come from the Energy sector, but rather from the combination of Agriculture and Changes in the Use of Soils and Forestry (CUTS in Spanish). Agricultural and forestry activities together are responsible for 31% of global emissions of greenhouse gases (GEG). In Latin America this proportion reaches 63%, which shows the importance of these sectors in the matrix of emissions in the region (see Table 1 on page 98).

The burden of deforestation (the reason for the positive balance in CUTS) in Brazil is huge, reaching 818 million tons of CO_2 emissions, a similar number to the total emissions reached by burning of fossil fuels in Germany (808 million tons CO_2 in 2004, according to UNDP).[4] This is very important for two reasons: firstly, because it highlights the importance of deforestation as a contributing factor of the region towards climate change; and secondly, because it shows that avoiding deforestation is a measure that may have positive impacts at various levels.

This means that although the contribution to climate change as a sub-region may not be very important, in relative terms, the contribution of Latin America is closer to that of developed countries than to that of developing countries.

This does not only have implications on the climate. Latin America has between 18% and 26% of the world's carbon sinks in their forests, apart from grasslands and agricultural ecosystems. However, these are being rapidly lost as a consequence of agricultural and mining activities or the construction of infrastructure with very high environmental impacts, far beyond those that cause greenhouse effects.

The high levels of greenhouse gas emissions in non-energy sectors—particularly those derived from deforestation—make the per capita market averages of many countries in South America higher than the values of various developed countries. Latin America is not a minor contributor to climate change considering the emissions per capita or GDP per unit. With 8.5% of the population and global GDP, the region is responsible for 12% of global greenhouse gas emissions. Consequently, although its

---

4 UNDP, Human Development Report 2007/2008; Fighting climate change: Human solidarity in a divided world <http://hdr.undp.org/en/reports/global/hdr2007-2008> accessed 06 July 2012.

responsibility regarding climate change is minor—as it is an accumulation of gases produced mainly in developed countries during past decades—it is important for climate change in the future. Leading up to 2050, even assuming that industrialized countries will reduce their emissions to zero, developing countries must reduce their emissions by 28% with respect to the year 2000.[5] This would be a powerful reason for making an effort to reduce greenhouse gas emissions.

3. Conclusions

There are various compelling reasons to confront a 'climate economy' based on the economic, social and environmental benefits that a strategy of this kind would bring for the countries in the region. There are a number of mitigation measures recommended by the majority of public and private international organizations that are part of the historical needs in Latin American countries, and at present have more potential to be fulfilled than in the past. These measures are being encouraged mainly for reasons linked to climate change, but the countries in the region must take advantage of these new options due to the additional benefits these measures imply.[6]

4. The Sense of Urgency

It seems that developing countries are not investing in mitigation and adaptation, awaiting the results of international debates on the responsibility of industrialized countries, postponing their options for development and waiting to get financial resources derived from 'ecological debts'. International financial resources such as CDM, the Adaptation Fund and others, are at present—and probably in the future—insufficient with regard to the needs of Latin American countries. There are no compelling reasons to wait for these, even though the claims are justified. However, they can already take advantage of the development of new markets and products offered, adding to the economic, social and environmental benefits that these changes propitiate, for the benefit of the Latin American society.

---

[5] Augusto De la Torre, A.; Pablo Fajnzylber and John Nash, *Low Carbon, High Growth: Latin American Responses to Climate Change: An Overview* (World Bank 2009).

[6] OECD (ed), Policy Brief; Cost Effective Actions to Tackle Climate Change, 2009.

The urgency of certain measures and the imperative of a more equitable and sustainable development should be a good reason to undertake actions in pursuit of a 'climate economy' as soon as possible.

Table 1. Greenhouse Gas Emissions in South America

Numbers in Millions of tons of CO_2	Energy	Agriculture	CUTS	Others	Total
Argentina	127	115	-34	21	229
Bolivia	9	9	26	3	47
Brazil	248	369	818	42	1477
Colombia	62	61	14	15	152
Ecuador	19	8	14	4	45
Mexico	321	39	141	23	524
Peru	22	22	41	13	98
Total ALyC	943	961	928	156	2986

The energy sector was the principal source of emissions in all regions, except in Latin America and the Caribbean, where agriculture occupied the first place (....) The LULUCF sector is the net sump of GEG in the region of Africa (411 million tons of CO_2 equivalent), in the region of Asia and the Pacific (316 million tons of CO_2 equivalent) and in 'other countries' (5 million tons of CO_2 equivalent).[7]

[7] UNFCCC (ed), Subsidiary Body for Implementation, Sixth compilation and synthesis of initial national communications from Parties not included in Annex I to the Convention, FCCC/SBI/2005/18/Add.2 of 25 October 2005.

CLIMATE PROTECTION AND SUSTAINABILITY IN JAPAN—TASKS FOLLOWING MARCH 11TH 2011

Masanori Okada[1]

1. Introduction

A disastrous earthquake and tsunami struck the east regions in Japan on March 11th 2011. The following day, enormous damage was caused by severe accidents of the reactors of the Fukushima nuclear power plant.

First and foremost, I would like to thank you all very much for your kind words of encouragement and the various efforts that have been made in order to help Japan!

In this short presentation I would like to comment on the challenges that Japan now faces with regard to climate protection.

2. National Politics under Frameworks of International Law

Being a party to the Kyoto Protocol (1997), Japan has committed itself to reducing its greenhouse gas emissions by 25% until 2020—and by 80% until 2050—against 1990 levels. According to the Japanese Government's plans these targets could and should be met by energy savings in all sectors, further promoting nuclear and renewable energies, implementing a carbon dioxide tax, carbon emissions trading and promoting innovation in the field of new technologies. Can Japan still, after March 11th, keep this scenario?

Many Japanese certainly do understand the importance of climate protection, but at the same time they wish for economic growth and employment.[2] Is there a way to link those two ideas without conflict?

---

[1] Professor of Law, Waseda Law School, Waseda University (Japan).

[2] Cf. WBGU (ed), Welt im Wandel: Gesellschaftsvertrag für eine Große Transformation, 2011, p. 75f. <http://www.wbgu.de/fileadmin/templates/dateien/veroeffentlichungen/hauptgutachten/jg2011/wbgu_jg2011_ZfE.pdf> accessed 15 July 2012.

3. A Critical Appraisal

3.1. *Industrial Metabolism*

At times, leaders of the industrial sector claim that any climate protection measure that will hinder economic growth should not be taken. But this approach ignores the cost of negative effects of future global climate change and overestimates the cost of implementing climate protection measures.

3.2. *Over- and Underestimating New Technologies*

So far in Japan, while the role of nuclear power plants as a means of preventing global climate change and the possibilities of controlling the risks of this technology are being overestimated, environmentally friendly energies such as solar power are being underestimated. 'Renewable energies are unreliable and expensive' is one common example for how people see this technology.

3.3. *Dependency on Nuclear Power Plants*

Over the last 30 years the quota of energy being produced by nuclear power plants in Japan has increased year after year (it now amounts to around 12.5% of nuclear energy worldwide and to 30% of national Japanese electricity generation). The scenario depicted by the Japanese government implies that 5% of the goal of reducing CO_2-emissions by 25% until 2020 have to be achieved by using nuclear power plants. Knowing this, nine leading Japanese electric power industries came to an agreement with the government and bureaucrats: They are allowed to generate a surplus by adding the cost of nuclear power plants to the general electricity rate. This is possible due to the regional monopoly of said electric power industries and the questionable cost calculation of electricity generation.

3.4. *Climate Protection Measures as a Political Steering Device*

Prior to March 11th, climate protection as a keyword did not include plans to overcome the status quo in the field of nuclear power generation. This was due to the government and the nuclear power lobby promoting nuclear power as an important contributor to climate protection. In my opinion there were three real reasons for this governmental strategy: First, the construction of nuclear power plants is a financial measure against farming and fishing areas, secondly, the Japanese government wants to show the industrial and military presence of Japanese nuclear technology, and thirdly, it must promote close cooperation with the U.S. military strategy to control Japan's military technologies and at the same time reduce public rejection of nuclear weapons in Japan.

4. Rebuilding and Climate Protection

Following the natural disaster and nuclear catastrophe at Fukushima power plant, Japan now has to find new ways of protecting the climate. Some academic societies have already made proposals for a structural change in the industry and in the way of life. The Japanese government toyed with the idea of revoking the 25% reduction target, but in the face of opposition now focuses on different solutions. For instance, there are plans for a law promoting renewable energies, which includes the initiation of a system making the buying of renewable energy at a fixed price mandatory.

5. Conclusion

I hope that we are going to make a virtue out of necessity by gaining a new perspective on sustainability and global public interest through the discussion on climate protection in this changed situation.

6. Supplementary Remarks

COP 17, held in Durban, adopted the 'Durban Agreements', which extend the Kyoto Protocol and start fresh negotiations on a new international framework for cutting global greenhouse gas emissions. Though Japan's Minister of Environment stated in the session of this Conference that there is no change to Japan's stance of tackling climate change even after the national crisis following the East Japan Great Earthquake, Japan, along with Canada and Russia, decided not to take part in the Kyoto Protocol's second commitment period.

This decision of the Japanese government is in contradiction to the 'Manifesto' of the government party, the Democratic Party of Japan. Many academic societies and environmental organizations criticize this change of international environmental policy and propose that Japan should contribute to the realization of a 'Low-carbon society after the Accident of Fukushima Daiichi Nuclear Power Plant' in international society, including through participation in the second commitment period.

DISCUSSION SUMMARY
WORKING GROUP: INITIATIVES ON AN EU AND ON A NATIONAL LEVEL

Astrid Epiney[1]

The discussion in the working group has illustrated and shown different initiatives on national levels, but also on the EU level, in order to contribute to climate policy. The situation in the different countries is, however, rather different so that a real comparison between the different instruments and mechanisms presented does not seem possible nor useful in our context. At least some instruments have to be evaluated very carefully in both a regional context and in the context of other (possibly) lacking or insufficient instruments; so, e.g., the accent that European Union law puts on the ETS does not seem—for different reasons—appropriate. Also for an evaluation of the other instruments presented, one may refer to the written statements.

However, four overall conclusions may be drawn from the presentations and the discussion:

1. If it is certainly true that international commitments on different levels are not sufficient for a real effective decreasing of greenhouse gas emissions, this may not be advanced as a reason for the States to renounce on climate policy initiatives and instruments. On the contrary, perhaps the present difficulties to find a consensus on an international level may be a supplementary reason to push national initiatives. So, the lack of international commitments may be (at least partially) compensated by national initiatives. Furthermore, national initiatives may incite and / or accelerate developments on an international level.
2. In any case, climate policy asks for different instruments. Indeed, good reasons plead for the fact that only instruments in different sectors and using different mechanisms may be sufficient. Furthermore, they also may be complementary.
3. If it is true that climate policy has a price, it is nowadays established that the costs, especially for adaptation, will be even higher if there is

[1] Prof. Dr. Astrid Epiney, Université de Fribourg, Switzerland.

no effective climate policy. This is true for developing and developed countries and should be a reason for all countries to decide on effective instruments in the area of climate policy.

4. Especially for developing countries, there is a real need to establish mechanisms which may assure low carbon based development. It cannot be an alternative, neither for developing nor for developed countries, to have as a point of departure that developing countries have the same 'right' to grow on a high carbon based development as developed countries. However, this means also that developed countries have a special responsibility to contribute to enhance this goal.

PART THREE

ENVIRONMENTAL HAZARDS RELATED TO SHIPPING

Presentations

INTEGRATING SHIPPING INTO THE EU EMISSIONS TRADING SCHEME?

Tim Bäuerle[1]

1. Introduction

Although the actors of global climate policy have had and still have considerable difficulties to come to an agreement on how the Post-Kyoto era should be created in order to keep global warming under control, at least on the political level there has been some notable progress: At the climate conference in Cancún in 2010 the Parties of the UNFCCC decided 'to hold the increase in global average temperature below 2°C above pre-industrial levels'.[2] At the same conference, the European Union pledged to reduce unilaterally its emissions by at least 20% of 1990 levels by 2020. The European Union further offered to reduce emissions to 30% by 2020, on the condition that other major emitting countries commit to fair emission reductions in a future global climate agreement.

Although emission reductions from international aviation and international shipping are not accountable under the national emissions registries of CO_2-equivalent emissions set up under the Kyoto-Protocol, the European Union has taken all efforts to include the emissions of international transport into the international emission reduction commitments. The reasons for these efforts are obvious: All models predict considerable growth of emissions from the aviation and shipping sectors.[3] In case of emissions from aviation, it is long known that the upper layers of the atmosphere react much more sensitively to aviation emissions and that the radiative forcing of emissions from aviation is therefore two- to fourfold of the radiative forcing caused by land-based CO_2-emissions.[4] With

---

[1] LL.M., Rechtsanwalt, Heidelberg.

[2] Report of the Conference of the Parties on its sixteenth session, Decision 1/CP 16, FCCC/CP/2010/7/Add.1, I 4. 15 March 2011.

[3] Business-as-usual projections for 2020 are about 0.6 to 1.2 $GtCO_2$ per year from aviation and 1.1 to 1.3 $GtCO_2$ per year from shipping, UNEP, 'Bridging the Emissions GAP' (November 2011) 11. <www.unep.org/publications/ebooks/bridgingemissionsgap/> accessed 28 February 2012.

[4] IPCC, 'Aviation and the Global Atmosphere' (1999) 8<www.ipcc.ch/ipccreports/sres/aviation/index.php?idp=8> accessed 28 February 2012; WGBU—German Advisory Council

respect to emissions from shipping it is generally accepted that the negative environmental impact of the bunker fuels used in international shipping is more than considerable. Not only CO_2-emissions, but also black carbon,[5] SO_2 and NO_x-Emissions from international shipping have serious negative impacts on the atmosphere, the marine environment as well as on terrestrial ecosystems and the human health.[6] With respect to CO_2-emissions, various experts have shown that the reduction of emissions can be reached at very low—even negative—costs.[7] And finally policy makers of the European Union are interested to avoid any distortion of competition in the various sectors of the economy or in different modes of transport. From a European perspective, the inclusion of emissions from international transport largely caused by or in the interest of European citizens is also a question of equity.

Although the European Union increased the pressure on the negotiations that according to the UNFCCC should have begun immediately after the convention came into force, the results so far remained unsatisfactory. This is true for the negotiations at ICAO as well as for the results that could be achieved at IMO. In 2008, the European Parliament and European Council enacted Directive 2008/101/EC in order to include aviation activities in the scheme for greenhouse gas emission allowance trading within the Community. One year later, as a consequence of the difficult negotiations at IMO, the European Parliament and the European Council in 2009 took the following common decision in order to put pressure on IMO and its member states:

---

on Global Change, 'Charging the Use of the Global Commons' (Special Report 2002) 7 <www.wbgu.de/en/special-reports/sr-2002-charging-the-use/> accessed 28 February 2012.

5 Black carbon emissions from ships alone are thought to be responsible for up to 60.000 premature cardiopulmonary and lung cancer deaths annually, DLR, 'SeaKlim (Einfluss von Schiffsemissionen auf Atmosphäre und Klima)' (Final Report, 2011) <www.dlr.de/Portaldata/1/Resources/standorte/oberpfaffenhofen/dokumente/SeaKLIM_Nac.hwuchsgruppe_Final_Report.pdf> accessed 19 April 2012. The German NABU (Nature And Biodiversity Conservation Union) has helped to inform the public about the serious impacts of the use of bunker fuels in international shipping with a campaign that criticises the use of bunker fuel—especially in cruisers—and has gained considerable public attention <www.nabu.de/themen/verkehr/schifffahrt/mirstinkts> visited 19 April 2012.

6 BMT, 'Study on the Economic, Legal and Practical Implications of a European Union System to Reduce Ship Emissions of SO_2 and NO_x' (2000) <http://ec.europa.eu/environment/enveco/taxation/ship_emissions/pdf/mainfinal.pdf> accessed 19 April 2012.

7 IMO, 'Second IMO Greenhouse Gas Study' (2009) 199 <www.imo.org/blast/blastDataHelper.asp?data_id=27795&filename=GHGStudyFINAL.pdf> accessed 28 February 2012.

> Developed countries, including the EU Member States, should continue to take the lead by committing to collectively reducing their emissions of greenhouse gases in the order of 30% by 2020 compared to 1990. [...] In the event that no international agreement which includes international maritime emissions in its reduction targets through the International Maritime Organisation has been approved by the Member States or no such agreement through the UNFCCC has been approved by the Community by 31 December 2011, the Commission should make a proposal to include international maritime emissions in the Community reduction commitment with the aim of the proposed act entering into force by 2013. Such a proposal should minimize any negative impact on the Community's competitiveness while taking into account the potential environmental benefits.[8]

So, whereas the European Union has taken action in the field of aviation, political action with respect to CO_2-emissions of international shipping is still pending. This situation gives the chance to take into account the experiences that so far have been made with the inclusion of emissions from aviation into the European Emission Trading System. However, the question of whether the European Union will take action in the field of maritime emissions can probably still be decided at IMO.

2. Political Progress and Backlashes at IMO

The efforts at IMO to take common action with respect to a regulation of greenhouse gases from international shipping go back to 1997. It took two studies on greenhouse gases from shipping to bring serious momentum into the negotiations on abatement measures. The work at the Marine Environment Protection Committee (MPEC) was soon divided into three elements: the technical, the operational and the market based measures. Under pressure from the European Union, IMO achieved a breakthrough at the 62th session of the MPEC in July 2011. The Committee adopted amendments to MARPOL Annex VI for the inclusion of a binding Energy Efficiency Design Index that should become binding for all merchant ships of 400 gross tonnage or above and will require new ships to be at least 10% more energy efficient from 2015, 20% more efficient from 2020 and 30% from 2025. Additionally MPEC decided to make mandatory a Ship Energy

[8] Decision 406/2009/EC of the European Parliament and of the Council of 23 April 2009 on the effort of Member States to reduce their greenhouse gas emissions to meet the Community's greenhouse gas emission reduction commitments up to 2020 [2009] OJ L 140/136.

Efficiency Management Plan (SEEMP) for all ships in action. These results of MPEC 62 shall come into force from the beginning of 2013.[9] However, in the working field of market based measures negotiations were once more adjourned at MPEC 62. Whether the Parties to IMO can come to a common understanding with respect to the introduction of an international regime of market based measures in order to limit the strong increase of emissions from shipping as prognosticated by the second greenhouse gas study[10] remains far from certain, although the secretary and at least part of the Member States openly acknowledge that a binding Energy Efficiency Design Index (EEDI) is not sufficient in order to regulate the greenhouse gas emissions of international maritime shipping. There are at least three strong arguments for additional market based measures: First, the EEDI will only cover new ships and therefore has no impact on vessels already on sea. As the investment circle for vessels is relatively long, it will take up to thirty years until all ships on sea have adapted to an EEDI that by then will be no longer state of the art. Second, all studies forecast a very strong growth in maritime shipping, that will offset the positive effects of an obligatory EEDI. And, probably most important, it is well known that the costs for a reduction of greenhouse gases from maritime shipping are quite low, in many cases even negative. All recent studies show that the growth of emissions could be reduced or even reversed considerably if all measures with negative costs would be applied.[11]

Considering the long time it took to reach an agreement on a binding EEDI at IMO and given the ambitious aims the European Union set itself with respect to the reduction of greenhouse gas emissions in the maritime sector,[12] it is more than understandable that the European Union as international actor has the feeling that—even if taking an optimistic view—it may take some time until the Member States of IMO take measures that regulate international shipping in a way that shipping bears an adequate

---

9 <www.imo.org/ourwork/environment/pollutionprevention/airpollution/pages/breakthrought-at-mepc-62.aspx> accessed 20 April 2012.

10 IMO (n. 7).

11 The different results of the Studies reflect different assumptions on the development of bunker fuel prices and the costs of technical abatement measures.

12 The Commission has declared that the GHG emissions from international shipping can be reduced by 40% by 2050 (compared to 2005 levels), European Commission, 'White Paper on Transport' (2011) <http://eur-lex.europa.eu/LexUriServ/LexUriServ.do?uri=COM:2011:0144:FIN:EN:PDF> accessed 20 April 2012. The European Council set as a target a reduction of 20% by 2020 compared to 2005 levels (Environmental Council Conclusion 2009), Council Conclusions on EU position for the Copenhagen Climate Conference (7–18 December 2009), 2968th Environment Council meeting, 21 October 2009, <http://www.consilium.europa.eu/uedocs/cms_data/docs/pressdata/en/envir/110634.pdf> accessed 20 April 2012.

	2005	2020	2030	
Source		IMO 2009	DNV 2010	CE Delft
Maximum abatement potential of all measures	955 Mt	−10.5% to +8.9%	−30%	−29.7% to +10.6%
Maximum abatement potential of measures whose cost efficiency is negative		−7.3% to +16.7%	+7.8%	−26.7% to +16.6%

Source: European Climate Change Programme—WG Ships: Background document: The level of reduction in the maritime sector.[13]

load part in the international efforts to reduce greenhouse gas emissions. Under these circumstances it remains very likely that—in line with Directive 2008/101/EC—the Commission will recommend taking unilateral action in order to regulate international maritime shipping from and to the European Union as a first step.

3. Actual Activities of the European Commission

Following Directive 2008/101/EC, the Commission had to observe that by the end of 2011 a binding agreement had neither been approved of at IMO nor within the UNFCCC. Therefore, the European Commission was obliged to take action with a view to initiate adequate measures with an aim to develop a proposal to include international maritime emissions in the Community reduction commitment as indicated in the Directive:

- Under the ECCP II Programme (European Climate Change Programme II) the Commission established a special working group (Working Group 6—Reducing Greenhouse Gas Emissions from Ships). The working group shall assess and develop measures to reduce greenhouse gas

[13] The background document of the ECCP—WG 6 is available under: <http://ec.europa.eu/clima/events/0047/level_reduction_en.pdf> accessed 19 April 2012. The background document summarizes important results of the major studies recently done in the field of greenhouse gas emission reduction, CE Delft et al., 'Technical Support for European Action to reducing Greenhouse Gas Emissions from International Maritime Transport' (2009) <http://ec.europa.eu/clima/policies/transport/shipping/docs/ghg_ships_report_en.pdf> accessed 19 April 2012; IMO (n. 7) and DNV (Det Norske Veritas AS), 'Pathways to low carbon shipping—abatement potential towards 2030' (2010) <www.dnv.com/binaries/pathways_to_low_carbon_shipping_tcm4-420472_tcm4-428628.pdf> accessed 19 April 2012.

emissions from shipping and to mainstream such measures into the European Climate Policy.[14]

- In parallel, the Commission started consultations within the different General Directions by establishing a High Level Platform with the competent commissioners, Vice President Siim Kallas (DG Transport) and Commissioner Hedegaard (DG Climate Action).[15]
- In January 2012, the Commission started an online-consultation on possible measures to reduce greenhouse gas emissions from shipping.[16]

As a result of the online-consultation it is very likely that the Commission will decide which kind of policy it will take in order to include maritime emissions into the European climate policy.

In the ECCP, Commission and participants already identified some possible measures. At the last meeting of the EECP (WG 6) the Commission provided a background paper on the 'main possible policy actions' that identified four main measures that could bring the results the Commission is aiming at.[17]

3.1. *Levy / Compensation Fund*

This policy option would aim at a levy on fuel for the maritime industry. Two possible models of this option are in discussion: One concept is designed as straight levy on fuel, the other concept is designed as a 'voluntary' fund managed by the shipping industry, but coupled with a tax of punitive character for those ships that do not take part in the 'voluntary' fund. In both concepts the proceeds of the levy would go to a compensation fund that has the objective of financing measures for the emission reduction in the shipping industry. The design of this measure follows an industry managed NO_x- fund that has been set up by Norway with some success.[18]

---

[14] <http://ec.europa.eu/clima/policies/eccp/second/stakeholder/index_en.htm> accessed 19 April 2012.

[15] <http://ec.europa.eu/clima/events/0037/index_en.htm> accessed 19 April 2012.

[16] <http://ec.europa.eu/clima/news/articles/news_2012011901_en.htm> accessed 19 April 2012.

[17] It must be pointed out that the Commission emphasizes that the decision making process within the Commission is still open and that the background document does not indicate any preference of the Commission. European Climate Change Programme—WG Ships, 15–16 November 2011, Background Document: Main possible policy options <http://ec.europa.eu/clima/events/0047/policy_options_en.pdf> accessed 19 April 2012. The following overview is based on the background document of the Commission.

[18] Ibid.

3.2. *Emission Trading Scheme*

Considering the policy the Commission pursued in the field of aviation, it is obvious that inclusion of the maritime transport sector into the European emissions trading systems is one of the policy options under discussion. However, due to the characteristics of the maritime industry, the details of an emission trading scheme that covers greenhouse gas emissions from vessels would pose some difficult decisions:

- Which entity should be responsible for providing the necessary allowances (ship owner, ship operator or ship manager)?
- Which emissions should be covered by a trading scheme (based on the actual last trip(s) or a certain period of time)?
- Which vessels should be covered (only very large vessels)?
- Should the system be open, semi-open or closed vis-à-vis the existing ETS, i.e. should allowances from other sectors be used for shipping (and vice versa)?
- How should allowances be distributed (auction vs. grand fathering)?[19]

3.3. *Taxation on Fuel or Greenhouse Gas Emissions*

Another option is the taxation of fuels for maritime shipping or on greenhouse gas emissions of ships. Such a tax could be fuel-based or emission-based. In the case of a fuel-based tax it would be likely that the fuel-supplier would be the payer of the tax. For an emission-based tax, it would be more likely that the take is made payable by the vessel as legal entity (i.e. ship owner, ship operator or ship-manager).[20]

3.4. *Mandatory Ship-level Emissions Reduction*

The last proposal under examination aims at a mandatory emissions reduction on ship-level. That would mean that every vessel visiting European harbours would come under an obligation to fulfill a special emissions reduction plan. This plan would probably depend on certain vessel characteristics, e.g. age, type of ship, emissions etc. The proposal could be

---

[19] These questions have been discussed in some detail in a study commissioned by the German Ministry of Environment: Öko-Institut, Margareta Kulessa, Matthias Oschinski and Tim Bäuerle, 'Integration of Marine Transport into the European Emissions Trading System' (2010) <www.umweltdaten.de/publikationen/fpdf-l/3942.pdf> accessed 19 April 2012.

[20] European Commission (n. 17).

coupled with an incentive system for 'overachievers', who then could sell additional abatement achievements on a market provided by the Commission or the Member States.

3.5. *Other Measures under Discussion*

Most other measures under discussion no longer seem to play a decisive role in the deliberations of the Commission. This is especially true for proposals that aim at a direct regulation of the speed of vessels.[21] It probably soon became clear that an indirect regulation (by a fuel tax, a levy or an ETS) would be more efficient. Furthermore, speed regulation would pose serious legal problems in terms of prescriptive and enforcement legislation for the European Union. It would be very difficult to argue that compliance to the regulation of shipping speed in international waters can be understood as a valid condition to entry into EU-ports.

3.6. *Relevant Legal Aspects in the Decision-Making Process of the Commission*

Apart from technical and administrative aspects there are at least two legal aspects that will influence the outcome of the decision-making process of the Commission: One will be the internal legislative procedure within the European Union and especially the question whether a certain measure will require unanimity or whether legislation could be enacted with majority voting according to Art. 205 II TFEU, whereas the introduction of a tax or a levy would require unanimity of all Member States.[22]

The second legal aspect influencing the outcome of the decision making process of the Commission will be the compatibility of the measure planned with international law.

---

[21] See for example the proposal by Sea at Risks, 'Speed Limits for Ship—Possible EU Options for tackling GHG Emissions from Ships' (ECCP Ship Working Group 2, 22 June 2011) <http://ec.europa.eu/clima/events/0036/speed_en.pdf> accessed 19 April 2012.

[22] The differences in the legislative procedure were already taken into account by the Commission when drafting the Background paper on main possible policy options (n. 17).

4. Learning from Aviation—The Judgement of the ECJ (Case 366/10)

In 2008, the European Parliament and the council amended directive 2003/87/EC so as to include aviation activities in the existing emission trading system of the European Union ('EU-ETS'). Opponents to the inclusion of aviation into the EU-ETS always held that the UNFCCC had transferred the matter of a regulation of emissions from international aviation to ICAO. Furthermore the opponents brought forward that the inclusion of emissions into the EU-ETS that are caused by planes outside the territory of the Member States of the European Union was not lawful. Similar objections have already brought forward against any unilateral action of the European Union with respect to the regulation of GHG-emissions from international shipping. When deciding on action in the shipping sector, the Commission therefore will take into account the experiences that have been found in the aviation sector.

4.1. *Legal Arguments against Measures in European Ports*

There are two main arguments against unilateral measures by the European Union for the reduction of greenhouse gases, both in the field of aviation as well as in the field of maritime transport. One argument is that any action that is aimed at the regulation of greenhouse gases emitted outside the territory of the European Union amounts to unlawful extraterritorial action. Other objectors point out, that under the UNFCCC all activities related to greenhouse gas emissions in the field of aviation and maritime transport have been referred to the competent international organisations i.e. ICAO and IMO, and this decision taken in the UNFCCC should be understood to be binding on the Member States of the UNFCCC. The judgment of the European Court of Justice on aviation has dismissed both arguments in very clear language.[23]

4.1.1. *Unilateral Regulation in Ports Amounts to Unlawful Extraterritorial Action*

Some legal observers have raised concerns that any action on side of single states that aims at emission reduction outside the territory of that state

[23] ECJ Case C-366/10 *The Air Transport Association of America and others.*

would amount to an unlawful exercise of extraterritorial jurisdiction.[24] Other commentators have been more optimistic that such action, notwithstanding its extraterritorial character, could be justified under international law.[25] The European Court of Justice however, somewhat surprising, holds on this question:

> 124. On the other hand, European Union legislation may be applied to an aircraft operator when its aircraft is in the territory of one of the Member States and, more specifically, on an aerodrome situated in such territory, since, in such a case, that aircraft is subject to the unlimited jurisdiction of that Member State and the European Union.....
>
> 125.... [The Directive] does not infringe the principle of territoriality or the sovereignty of third States from or to which such flights are performed have over the airspace above their territory, since those aircraft are physically in the territory of on the of the Member States of the European Union and are thus subject to the unlimited jurisdiction of the European Union.

As it is widely acknowledged that States have the same unlimited jurisdiction over vessels that are physically in a port of a certain State, the decision of the European Court of Justice can be transferred directly to the shipping sector: All market based measures actually under discussion within the Commission can be based directly on the territoriality principle.[26]

4.1.2. *The Kyoto Protocol Gives IMO (Exclusive) Competence to Regulate International Maritime Emissions*

Although the decision of the Court of Justice itself remains silent on the question, whether the provisions in the Kyoto Protocol hinder unilateral

---

[24] For example, Philip Roche and Emma Humpries from Norton Rose LL.P., London, have pointed out the risks of any extraterritorial action of the European Union with respect to measures aimed at a reduction of greenhouse gas emissions on the High Sea or in the EEZ of third countries, even if such measures are applied within the ports of the European Union. See C.E. Delft, 'Technical Support for European action to reducing Greenhouse Gas Emissions from international maritime transport' (2009) Annex B <www.cedelft.eu/publicatie/technical_support_for_european_action_to_reducing_greenhouse_gas_emissions_from_international__maritime_transport/1005> accessed 20 April 2012.

[25] See Henrik Ringbom, 'Global Problem—Regional Solution? International Law Reflections on an EU CO2 Emissions Trading Scheme for Ships' (2011) 26 The International Journal of Marine and Coastal Law 1, 20 et seq. and Öko-Institut, Margareta Kulessa, Matthias Oschinski and Tim Bäuerle (n. 19) 85 et seq.

[26] There remain doubts as to whether the judgement of the European Court of Justice is well founded in this point. As the acts that give rise to the regulation of the European Union (in part) takes place within the territory of the third States and the regulation of these acts are the direct objective of the legislation of the European Union, there are strong arguments to classify such legislation as extraterritorial and justifiable, if balanced with the interests of the third states concerned.

activities of its Member States, the legal opinion of advocate general Kokott gives some indication on how that issue should be answered.[27] Advocate general Kokott points out that the efforts of the European Union to find an international solution within ICAO in no way are reduced by the legislation enacted. As UNFCCC expressly acknowledges, Member States should take national measures in order to reduce greenhouse gas emissions. Thus, Art. 2 (2) of the Kyoto-Protocol cannot be interpreted so as to give IMO or ICAO an exclusive mandate to regulate greenhouse gas emissions of shipping or aviation respectively. This means that the European Union is free to regulate international shipping—within the limits of the jurisdiction of its Member States—at least until IMO is taking adequate international action, but probably even if the European Union is of the opinion that stricter regulation of maritime activities is necessary.

5. Conclusion

Considering the judgment of the European Court of Justice in the Air Transport Association Case it is quite clear that the legal constraints of international law in the design of a scheme for the reduction of greenhouse gas emissions in the shipping sector are quite weak. This, however, does not mean that the Commission can act without considering the interest of third States that could be affected negatively by the respective legislation. Experiences in the aviation sector show that third States may react with unpredictable countermeasures on European legislation that they perceive as an interference into their internal affairs. The tools provided by international law for assessing the limits of extraterritorial jurisdiction therefore could prove to be helpful for the Commission when deciding on the right measures: By balancing the interests of the European Union with the known interests of third States in advance, diplomatic tensions can be reduced. This is especially the case if States that play an important role in international maritime transport are adequately included in the consulting process that leads to the final legislation.

[27] Art 2 (2) of the Kyoto-Protocol states: 'The Parties included in Annex I shall pursue limitation or reduction of emissions of greenhouse gases not controlled by the Montreal Protocol from aviation and marine bunker fuels, working through the International Civil Aviation Organisation and the International Maritime Organisation, respectively.'

A COOPERATIVE COMPLIANCE STRATEGY: THE VOLUNTARY IMO MEMBER STATE AUDIT SCHEME

Lawrence D. Barchue, Sr.[1]

1. Introduction

The need to improve the implementation and enforcement of the various international conventions relating to the design, operation and maintenance of the world's commercial shipping fleet was brought to the fore in the early nineteen nineties. No longer was there an absence of internationally agreed standards, following the decade of the eighties when major safety, pollution prevention and training conventions had entered into force; it was now time to promote and ensure compliance by all actors in the shipping industry with those standards. Although the conventions referred to above had entered into force internationally, shipping casualties and serious incidents were still too common, thus the credibility of the international regulatory regime for shipping was being challenged to do more.

To address the apparent lapse in compliance with international shipping conventions, the International Maritime Organization (IMO) had to consider options for establishing recognized accountability regimes for the various actors involved in shipping. At the forefront of this drive to improve compliance were the establishment of the Flag State Implementation (FSI) Sub-Committee of the IMO in 1992; the expansion of Port State Memorandum of Understanding (PSCMoU) from the Paris MoU as the sole PSC regime to eight additional PSCMoUs covering the entire globe; the development of the ISM Code and the revision of the 1978 STCW Convention, with the latter creating, for the first time, a degree of State accountability with respect to the requirements of that Convention.

---

[1] Senior Deputy Director, Member State Audit and Internal Oversight Services, International Maritime Organization (IMO), 4 Albert Embankment, London SE1 7SR, United Kingdom, e-mail: lbarchue@imo.org. The views expressed in this article are those of its author and should not be taken as reflecting in any way the policies or views of IMO or of its Secretariat.

The aforementioned measures were not sufficient and the absence of a broader regime for State accountability, with respect to their internationally agreed conventions' obligations and responsibilities, was viewed as the weak-link in enhancing further the implementation and enforcement of those conventions' standards. In this respect, a compliance monitoring regime of IMO Member States' implementation and enforcement of the provisions of the various mandatory instruments of IMO had to be considered and developed.

This article will examine some of the critical components associated with compliance regimes directed at the principal actors and their roles in international shipping. It will explore the rationale for the development of a compliance regime for States and how such a regime has been accepted as a cooperative compliance mechanism by the States.

2. Who are the Principal Actors, Their Roles and Compliance Regimes

Regulating, implementing and enforcing international standards on shipping involve a multitude of actors, who may be spread across several jurisdictions. Thus, developing an effective compliance regime to ensure that all actors perform as required by the various international conventions that regulate shipping is complicated. Notwithstanding this expensive global industry, which is the engine of world trade, there are five principal actors, whose roles and responsibilities are as follows:

- The IMO, as a specialized agency of the UN, has the responsibility to develop global technical safety, security and pollution prevention standards relating to ships and shipping activities.
- Governments/States have the duty to implement and enforce these standards.
- Recognized organizations (ROs) have a duty to be impartial and exercise due diligence in carrying out statutory tasks delegated to them by governments.
- Shipowners/shipping companies are responsible for the consistent application of the same standards to individual ships.
- Seafarers have the task of putting into operation the various safety, security and anti-pollution measures applicable to the ship.

Understanding the roles of these five principal actors and who is accountable to whom is important. The IMO is a representative body of governments that develops and establishes the international standards relating to shipping. Through this body, governments also agree to measures that bind them to carry out certain regulatory and enforcement functions on ships and on the other actors as well. However, States are sovereign and when the mix of actors involved in implementing and enforcing shipping standards is considered, coupled with multiple jurisdiction issues, any lapse in coherent governmental actions, either through the IMO or in meeting previously agreed obligations and responsibilities, shipping could become a global industry plague by regulations from multiple jurisdictions. It is precisely to eliminate such independent regulatory action by States, with the associated consequences, as well as to continuously develop mechanisms to retain confidence amongst nation States that international shipping is safe, secure and environmental sound, that IMO was established and has been most successful.

This success of IMO is primarily in the regulatory aspect of its work. However, the absence of an internationally established compliance verification mechanism, leaves the hard earned improvements in safety and environmental standards for ships in the hands of the individual actors. To highlight this point, one needs to briefly examine the state of play of the principal actors.

Under various treaties, governments are responsible for ensuring compliance by the other four actors with the provisions of such treaties. However, some of these same treaties provide unrestrained powers to States to delegate statutory work, which also entails compliance verification on other actors. They also provide additional latitude for States to determine their own shipping standards through the phrase 'to the satisfaction of the Administration' and equivalency and exemption provisions. As a result, national laws to implement international shipping treaties could vary considerably and this leads to:

- partial or full delegation of statutory work to non-State entities;
- different degrees of implementation and enforcement;
- ship registration becoming an attractive and legitimate business in the absence of State accountability; and
- some shipowners enjoying considerable economic advantage due to the lack of uniform flag State enforcement.

With respect to ROs, they are granted authority to act on behalf of flag States and, as such, they form part of the State's enforcement mechanism. A number of ROs also have other business interests with ships entitled to fly the flag of the State that has delegated authority to such ROs. This leads to the intensification of commercial pressures on the ROs concerned, which sometimes create conflict between the recognized organization's role as certifier and inspector on behalf of the flag State and their commercial relationship as the classification society for the ship. For those ROs that are also classification societies, the ultimate sanction for non-compliance by a ship with the classification rules is the loss of class. In the absence of regulated and mandatory transfer of class agreements, there is no mechanism to ensure that the 'gaining' society will take responsibility and action to ensure that any outstanding conditions or items left at the time the ship is transferred are rectified.

The inclusion of resolution A.739(18) ('Guidelines for the authorization of organizations acting on behalf of the Administration') and resolution A.789(19) ('Specifications on the survey and certification functions of recognized organizations acting on behalf of the Administration') as mandatory requirements in regulation 1 of SOLAS chapter XI-1 (*Special measures to enhance maritime safety*) has not fully produced the desired result. Parties to SOLAS periodically provide information to IMO on the authorization granted to ROs based on the aforementioned regulation; but there is no independent mechanism to verify that the provisions of both resolutions are scrupulously adhered to by Parties and ROs. As a consequence, the competence of the numerous ROs, their representatives and ROs accountability to the international community cannot be verified.

On the other side of the divide are shipowners, companies and ship managers. The change from 'traditional' shipowner to the syndicate or 'asset player', the increasing use of bareboat or demise charter, time charter, single voyage or spot charter and single ship companies have made it all the more difficult to establish the identity of the party ultimately responsible for the ship. Although the introduction of the ISM Code has consolidated ownership and management of a ship into the definition of the 'company', there is still some latitude in the regulatory and enforcement regime that allows the principal actors (governments, recognized organizations and companies) to remain fully unaccountable for their actions.

3. A New Approach to Compliance Monitoring in International Shipping

From an esoteric point of view (maritime affairs), sovereignty can be seen and argued from many different perspectives other than the well-known arguments referred to in the preceding paragraphs. Sovereignty is the shared acceptance amongst nation States that, within a defined geograhical boundary, the people therein have constituted and accepted a legal and social framework that will govern their internal behaviour and security, as well as their external interaction with other nation States. This 'gentlemen's agreement' has provided the guarantee that the rights and privileges of the people of a State will receive reciprocal acceptance and treatment by the peoples of other States. This may sound like a classic definition of what is the norm; however, in the context of maritime activities, particularly those involving the extension of such sovereignty to ships, there is a second part.

The freedom to trade through the use of ships goes back centuries and as such there are some unregulated practices that still continue today as a matter of tradition. In addition to the ship being the sovereign territory of the flag State, ships have always been allowed to load and unload cargoes in ports without hindrance. In saying so, ships are operated as if they have established rights to enter any port on short notice to carry out commercial activities. Contrary to this established and accepted practice, there is no legal requirement that a State must allow any ship to enter its internal waters and ports to carry out commercial activities. This can be applied equally to the humanitarian aspect involving a ship in distress and in need of assistance. In such a case a coastal State is under treaty obligations to render assistance; but that assistance does not necessarily mean that the ship has a right of entry into the port of that State.

The extension of sovereignty to a ship entitled to fly the flag of State is now largely guaranteed through international treaties. Parties to such treaties accept, *ipso facto*, that the other Party is conforming to the agreed rules of play. The treaties referred to are primarily those adopted through the IMO relating to international shipping. The famous clause contained in a number of such treaties, which refers to 'no more favourable treatment for non-Party ships' provides sufficient latitude for intervention on a non-Party ship when such a ship is within a State's defined and accepted geographical boundary. This is an accepted fact and ships have been denied trade entry rights to ports or have been subjected to thorough port State control intervention if the flag State is not a Party to one of the relevant treaties.

On the other hand, the assertion of sovereign rights by flag States to ships entitled to fly their flag in dealings amongst States that are Party to various treaties is a diminishing notion. One needs only to look at the trend in port State control interventions, legal recourse by coastal and other States relating to alleged pollution from ships, irrespective of where the pollution incident occurred, and interdiction agreements on the high seas of ships suspected of conveying illegal cargo or activities. Notwithstanding this trend, the issue of a ship being the sovereign territory of the flag State has not been totally removed and still commands some degree of respect amongst Parties to IMO treaties.

In the context of compliance monitoring, coastal and port States may not continue to accept that the foreign sovereign entity (the ship) can enter the sovereign jurisdiction of another State, with the ship enjoying the protection of its sovereignty provided for under various treaties, whilst its flag State remains unaccountable with respect to its responsibilities and obligations that are derived from the treaties concerned. It is this other aspect of sovereignty and the increasing trend of intervention by coastal and port States that could provide additional impetus for Member States to accept the need for a cooperative compliance monitoring regime.

The need for a recognized mechanism to constructively assess Member States compliance with their obligations cannot be over-emphasized. Such an assessment, as envisaged under the Audit Scheme, could provide a basis for maintaining the status quo of mutual acceptance and recognition of compliance, based on the successful audit of all Member States. This could stem any further erosion of the notion that the ship is the sovereign territory of the flag State, even when it is operating within the territory of another State. It would restore a certain degree of assurance that the ship, entitled to fly the flag of a Party to the relevant treaties, should first be given the benefit of the doubt that it is in compliance with the relevant standards.

A cooperative compliance monitoring regime would also verify how coastal States meet their obligations under the various treaties and promote further the safety of navigation in coastal waters and the protection of the marine environment, whilst at the same time it could increase local commitment to the provision of search and rescue services for both domestic and international users of the sea.

With respect to port States, there could be a more judicious exercise of their rights granted under international treaties, which would be commensurate with their responsibilities to both the ship, its crew and the associated flag State. Best practices in port State control could be readily accepted and those port State control activities that are below par could also be curtailed.

4. Conclusion

This article has discussed the roles of the principal actors in shipping, their relationship and participation in the regulatory process and the need to establish a robust regime of accountability for all actors in shipping. Some components of such a regime already exist, such as the ISM Code and the STCW White List process. However, for these to be fully effective, a mandatory State accountability regime with respect to maritime treaties' obligations should be promoted. A mandatory Audit Scheme, under the aegis of IMO, will serve international shipping very well and would enhance the credibility amongst States that governmental implementation and enforcement actions are effective and are carried out by all Parties to the various treaties.

THE 'ERIKA III' PACKAGE: PROGRESS OR BREACH OF INTERNATIONAL LAW?

Alexander Proelss[1]

1. Introduction

The 'Erika III' package is the third set of legislative measures with which the European Union (EU) has responded to the sinking of the oil tankers *Erika* (1999) and *Prestige* (2002).[2] It contains rules that are primarily intended to increase safety at sea—as opposed to security at sea. In a 2004 resolution on improving safety at sea, the European Parliament emphasized that 'rapid and complete introduction and strict enforcement' of the measures adopted by the Member States with the 'Erika I' and 'Erika II' packages must have top priority.[3] At the same, however, it made manifest that the Commission should quickly present its more comprehensive proposal for improving maritime safety as had already been announced.[4] With regard to the relationship between the International

[1] Alexander Proelss is Professor of Public Law, in particular International Law and European Union Law, and Director of the Institute of Environmental and Technology Law at the University of Trier.

[2] The 'Erika I' package consists of the following measures: Regulation (EC) 417/2002 of the European Parliament and of the Council of 18 February 2002 on the accelerated phasing-in of double hull or equivalent design requirements for single hull oil tankers and repealing Council Regulation (EC) 2978/94 [2002] OJ L64/1; Directive 2001/106/EC of the European Parliament and of the Council of 19 December 2001 amending Council Directive 95/21/EC concerning the enforcement, in respect of shipping using Community ports and sailing in the waters under the jurisdiction of the Member States, of international standards for ship safety, pollution prevention and shipboard living and working conditions [2002] OJ L19/17; Directive 2001/105/EC of the European Parliament and of the Council of 19 December 2001 amending Council Directive 94/57/EC on common rules and standards for ship inspection and survey organisations and for the relevant activities of maritime administrations [2002] OJ L19/9. The 'Erika II' package contains the following legislative acts: Regulation (EC) 1406/2002 of the European Parliament and of the Council of 27 June 2002 establishing a European Maritime Safety Agency [2002] OJ L208/1; Directive 2002/59/EC of the European Parliament and of the Council of 27 June 2002 establishing a Community vessel traffic monitoring and information system and repealing Council Directive 93/75/EEC [2002] OJ L208/10.

[3] European Parliament Resolution 2003/2235 (INI) of 30 April 2004 on improving safety at sea [2003] OJ C 104E/730 para 8.

[4] Ibid., para 9.

Maritime Organization (IMO) and the EU, the parliament, while expressing its understanding for the concerns raised by the Secretary-General of the IMO over the unilateral and regional actions of States outside the IMO framework, stated that

> [unilateral] EU action [...] may sometimes be necessary in the interests of safety [...], moreover, that EU measures can act as a catalyst within the IMO, as in the case, for example, of the accelerated phasing-out of single-hulled tankers.[5]

With this, the European Parliament not only embraced the widely held view that the global safety standards adopted under the auspices of the IMO were insufficient—the first recital of the Decision emphasizes that 'various incidents in European waters have caused pollution since the Erika and Prestige disasters'—but, at the same time, presented an opportunity to shape the manner of cooperation between the two organizations. Indeed, the EU is predestined for the realization and, above all, the enforcement of the pertinent international standards due to its supranational character and economic power.[6] It is precisely this effort to resolve a potential jurisdictional conflict that is meant when the European Parliament speaks about the EU becoming a 'catalyst' for the IMO.

One year later in its Communication of 23 November 2005, the Commission presented its third catalog of maritime safety measures as called for by the Parliament.[7] In no way did it limit itself to the sole objective of improving safety at sea, however. The fundamental problem identified here was 'transit traffic, outside the jurisdiction of the Member States, involving high-risk vessels flying the flag of third countries'.[8] According to the Commission, the measures already adopted would need to be intensified in a manner that takes the importance of the shipping industry for European competitiveness into account.[9] The ongoing development of European maritime safety policy would represent a contribution toward 'strengthen[ing] the safety aspects of the integrated European maritime policy being developed',[10] which would include economic, social and

---

[5] Ibid., para 36.

[6] Wolfgang Graf Vitzthum, 'Schiffssicherheit: Die EG als potentieller Durchsetzungsdegen der IMO' (2002) 62 ZaöRV 163, 177.

[7] Communication from the Commission COM (2005) 585 final of 23 November 2005 Third package of legislative measures on maritime safety in the European Union.

[8] Ibid., 6.

[9] Ibid., 3.

[10] Ibid.

ecological concerns as well as security issues. This would ultimately involve achieving a balance between the 'conservation of resources and the improvement of competitiveness, long-term growth and employment in the maritime sector'.[11] Indirectly, the Commission also aimed at the creation of an open market for maritime transport services at the Union and international levels.[12] This explains the considerably greater emphasis placed on maritime law in the 'Erika III' package in comparison to its two predecessors. In addition, it has often been voiced that the EU should become a member of the IMO[13]—a wish that has yet to be satisfied and, due to the lack of revision to the IMO's founding instrument,[14] which does not provide for the membership of international organizations, also could not be satisfied so far.

The legislative package proposed by the Commission consisted of six directives and two regulations. After the drafts were in part significantly modified over the course of the legislative process, the final versions of the individual components of the 'Erika III' package entered into force in May 2009. They were based on the narrow and subject-specific competence norm contained in Art. 100 (2) of the Treaty on the Functioning of the European Union (TFEU)[15] which arguably enjoys priority over other competence norms with a more general character that could have also been considered.[16]

In the following text, an overview of the 'Erika III' measures will first be presented before proceeding to the current case law of the European Court of Justice (ECJ), focusing in particular on the example of the *Intertanko* decision and its relevance for assessing the EU's competences as an actor in international relations. This primarily concerns the existence and location of the direct and indirect boundaries on the actions of the

---

11 Ibid.

12 Clearly ibid., 4.

13 See also Recommendation SEC (2002) 381 final from the Commission to the Council of 9 April 2002 in order to authorise the Commission to open and conduct negotiations with the International Maritime Organization (IMO) on the conditions and arrangements for accession by the European Community.

14 Cf. Art. 4 of the Convention on the International Maritime Organization of 6 March 1948, 289 UNTS 3: 'Membership in the Organization shall be open to all States, subject to the provisions of Part III'.

15 Consolidated Version of the Treaty on the Functioning of the European Union [2010] OJ C 83/47.

16 See Alexander Proelss, *Meeresschutz im Völker- und Europarecht: Das Beispiel des Nordostatlantiks* (Duncker und Humblot 2004) 314 et seq. Following Martin Nettesheim, 'Horizontale Kompetenzkonflikte in der EG' [1993] EuR 243, 248.

Union drawn by international law, as well as their potential transgression. Thereafter, the insights made in this process will be applied to the question of whether the components of the 'Erika III' package are compatible with global maritime safety standards valid under international law. The contribution concludes with reflections on the relevance of the topic for the general development of international law.

2. The 'Erika III' Package: An Overview

The 'Erika III' measures, on the whole, aim at establishing more stringent standards in regard to controls, liability and insurance.[17]

In accordance with Directive 2009/21/EC on compliance with flag State requirements,[18] a safety inspection must always be performed before a vessel is granted the right to fly the flag of an EU Member State (cf. Art. 4). The Directive provides that the administrations of the Member States are subject to an IMO audit at least every seven years (cf. Art. 7); this involves a scheme to review the extent to which a State has satisfied the obligations specified in the binding IMO instruments to which it has become a party.[19] Within the EU, the performance of such audits is thus no longer at the discretion of the Member States. In addition, the Member States must develop, implement and update a quality management system, which must be certified according to international standards, for the operational aspects of its administrative activities in regard to flag State obligations by mid-2012 (cf. Art. 8). Irrespective of these requirements, the provisions of the Directive are far behind that of the Commission's original draft, which aimed at establishing a comprehensive and truly European shipping administration (including the introduction of a European flag).

---

17 See already the general expositions by Yves van der Mensbrugghe, 'Le paquet Erika III: Un bouquet varié de propositions concernant la sécurité de l'Union Européenne', [2006] ADM XI 85 et seq.; id., Le paquet Erika III sur la sécurité maritime dans la Communauté Européenne enfin ficelé [2009] ADM XIV 295 et seq.; Uwe Jenisch, 'New EU Legislation on Safety at Sea 2009', (2009) 146 Hansa International Maritime Journal 52 et seq.

18 Directive 2009/21/EC of the European Parliament and of the Council of 23 April 2009 on compliance with flag State requirements [2009] OJ L131/132.

19 The following IMO documents form the basis for the audit scheme: A 23/Res.946 of 27 November 2003, Voluntary IMO Member State Audit Scheme; A 24/Res.974 of 1 December 2005, Framework and Procedure for the Voluntary IMO Member State Audit Scheme; A 24/Res.975 of 1 December 2005, Future Development of the Voluntary IMO Member State Audit Scheme.

Directive 2009/15/EC[20] and Regulation (EC) No 391/2009[21] concerning common rules and standards for ship inspections and survey organizations and for the relevant activities of maritime administrations replace and tighten the original Directive on classification organizations contained in the 'Erika I' package. They provide for the direct recognition and control of survey organizations by the Commission (cf. Art. 3 et seq. of Regulation [EC] No 391/2009) as well as the establishment of a 'working relationship' between the administrations of the Member States and the survey organizations working on their behalf (cf. Art. 5 of Directive 2009/15/EC), which should lead to clearer and more effective rules in regard to the performance of duties, liability and so forth.

The central rules on port State control were reformed by Directive 2009/16/EC.[22] This Directive provides for the introduction of a quality system for the most comprehensive possible control of all ships calling at ports in the EU on the basis of a more closely specified selection procedure. A risk profile must be created for every vessel calling at an EU port based on both generic (type of vessel, age, flag, classification organizations involved, and company performance) and historical parameters (number of deficiencies and detention measures) (cf. Art. 10). In regard to the basic inspection obligation and irrespective of the date of the previous inspection (cf. Art. 11), an expanded inspection procedure is applicable to vessels of the highest risk classification (cf. Art. 14). As *ultima ratio*, the Directive, which references the Paris Memorandum of Understanding (MoU) on Port State Control[23] in different areas (cf. only Art. 5, 8, 10), provides for temporary and ultimately permanent refusals of access (cf. Art. 16). This will be returned to as part of the analysis of the compatibility of the 'Erika III' measures with international law (see 4.3 below).

---

[20] Directive 2009/15/EC of the European Parliament and of the Council of 23 April 2009 on common rules and standards for ship inspection and survey organisations and for the relevant activities of maritime administrations [2009] OJ L131/47.

[21] Regulation (EC) 391/2009 of the European Parliament and of the Council of 23 April 2009 on common rules and standards for ship inspection and survey organizations [2009] OJ L131/11.

[22] Directive 2009/16/EC of the European Parliament and of the Council of 23 April 2009 on port State control [2009] OJ L131/57.

[23] The Paris MoU of 26 January 1982 is a non-legally binding agreement on the harmonization of the requirements for performing port State controls in Europe as well as Canada.

Directive 2009/17/EC[24] addresses amendments to the Directive on vessel traffic monitoring.[25] In the forefront here is the effort to establish a harmonized regime of places of refuge. The original approach intended—to authorize an independent European authority to designate a legally binding place of refuge to a vessel in distress—was abandoned due to opposition from Member States. In accordance with the Directive, the Member States themselves now designate one or more authorities, which shall be authorized to make independent decisions concerning the accommodation of ships in need of assistance (cf. Art. 20 of the revised Directive on vessel traffic monitoring). The proposed system of mandatory mutual financial guarantees and compensation payments ('maritime Eurobonds') for the benefit of States which open their ports to vessels in distress was also abandoned. Further revisions concern the strengthening of 'SafeSeaNet' as a platform for the exchange of information concerning the safety of specific vessels in European waters (cf. Art. 22 (a) of the revised Directive on vessel traffic monitoring), as well as the expansion of the compulsory automatic identification system for fishing vessels of more than 15m length (cf. Art. 6 (a) of the revised Directive on vessel traffic monitoring).

With Directive 2009/18/EC establishing the fundamental principles for the investigation of accidents in the maritime transport sector,[26] which builds on the IMO Code for the Investigation of Marine Casualties and Incidents,[27] technical rules for the investigation of maritime accidents in the transportation sector were introduced. A legal obligation to conduct a safety investigation exists only in the event of very serious maritime casualties or when substantial interests of a Member State are involved; the latter is seen to exist, for example, when the vessel flies the flag of this Member State, or when the accident occurs in its 'maritime aquitory'[28] (cf. Art. 5 (1)).

---

[24] Directive 2009/17/EC of the European Parliament and of the Council of 23 April 2009 amending Directive 2002/59/EC establishing a Community vessel traffic monitoring and information system [2009] OJ L131/101.

[25] This refers to Directive 2002/59/EC (n. 2) which was integrated into the 'Erika II' package at that time.

[26] Directive 2009/18/EC of the European Parliament and of the Council of 23 April 2009 establishing the fundamental principles governing the investigation of accidents in the maritime transport sector and amending Council Directive 1999/35/EC and Directive 2002/59/EC of the European Parliament and of the Council [2009] OJ L131/114.

[27] IMO Doc. A 20/Res.849 of 27 November 1997, Annex.

[28] This term introduced by Wolfgang Graf Vitzthum encompasses the maritime zones part of a State's territory (internal waters, territorial sea, archipelagic waters); it clarifies

The objective of Regulation (EC) No 392/2009 on the liability of carriers of passengers by sea[29] is the uniform implementation ('communitization') of the Athens Convention relating to the Carriage of Passengers and their Luggage by Sea of 1974 and its 2002 Protocol,[30] which had not previously been ratified by all EU Member States. It establishes a liability and insurance regime for the transport of passengers and their luggage. In this regard, the Directive goes beyond the Athens Convention as its rules also apply to domestic transport by sea (cf. Art. 2).

Finally, according to Directive 2009/20/EC on the insurance of shipowners for maritime claims,[31] Member States must require that the owners of ships with more than 300 gross tonnage flying their flags have insurance for these ships (cf. Art. 4). This insurance must cover maritime claims subject to the limitations of the 1976 Convention on Limitation of Liability for Maritime Claims in its revised version of 1996[32] (cf. Art. 4 (3)). The insured amount per vessel and incident corresponds to the maximum amount specified in the 1996 Convention. This constitutes an expansion and improvement to global shipping insurance law.

As already mentioned, the legal acts presented above were subject to numerous requests for revision by the European Parliament and the Council during the legislative process, which, at least in part, concerned their compatibility with global standards adopted under the auspices of the IMO. Also due to certain experiences with the 'Erika I' and 'Erika II' packages, the EU institutions seem to have shyed away from the European unilateralism considered appropriate by the Parliament in 2004. Before specific aspects of the 'Erika III' measures are examined against this background for their compatibility with international law, a glance will be taken at the experiences made with the preceding legislative packages.

---

that a State's sovereignty over its *territorium* can be qualitatively distinguished from that over its maritime territory. Cf. Wolfgang Graf Vitzthum, 'Aquitoriale Souveränität. Zum Rechtsstatus von Küstenmeer und Archipelgewässern', in Pierre-Marie Dupuy, Bardo Fassbender, Malcolm N. Shaw and Karl-Peter Sommermann (eds), *Völkerrecht als Wertordnung = common values in international law: Festschrift für Christian Tomuschat* (N.P. Engel Verlag 2006) 1067 et seq.

29 Regulation (EC) 392/2009 of the European Parliament and of the Council of 23 April 2009 on the liability of carriers of passengers by sea in the event of accidents [2009] OJ L131/24.

30 Consolidated version: [2009] OJ L131/29.

31 Directive 2009/20/EC of the European Parliament and of the Council of 23 April 2009 on the insurance of shipowners for maritime claims [2009] OJ L131/128.

32 Convention on Limitation of Liability for Maritime Claims of 19 November 1976, 16 ILM 606; Protocol Amending the Convention on Limitation of Liability for Maritime Claims of 2 May 1996, 35 ILM 1433.

3. Intertanko and the Limits of Union Competences in the Field of Maritime Safety

In the context of 'Erika I' and 'Erika II' as well as the measures adopted in the aftermath of the sinking of the single-hulled oil tanker *Prestige*,[33] it was above all in regard to the accelerated schedule for introducing a ban on single-hulled tankers stipulated by Regulation 1726/2003 as well as the stricter limits on sulphur in marine fuels established under Directive 2005/33 and their compatibility with the 1973 International Convention for the Prevention of Pollution From Ships (MARPOL)[34] where doubts had arisen.[35] Further points of contention included the extent of the obligations set out in Directive 2002/59 on vessel traffic monitoring.

3.1. *The ECJ's Intertanko Decision*

Notwithstanding this, it was questions concerning the provisions of Directive 2005/35/EC on ship-source pollution and on the introduction of penalties for infringements[36] and their compatibility with international law that finally landed before the ECJ in 2006. The plaintiffs in the initial proceedings (including the International Association of Independent Tanker Owners—Intertanko) claimed before the High Court of Justice of England and Wales that the introduction of the standard of 'serious negligence' for the investigation of liability for the discharge of polluting substances under the Directive, as well as its exclusion clause concerning the application of specific exemptions from liability, were not compatible with the rules contained in MARPOL Annexes I and II[37] and the UN Convention

---

[33] Included among these measures are Regulation (EC) 1726/2003 of the European Parliament and of the Council of 22 July 2003 amending Regulation (EC) 417/2002 on the accelerated phasing-in of double-hull or equivalent design requirements for single-hull oil tankers [2009] OJ L249/1, Directive 2005/35/EC of the European Parliament and of the Council of 7 September 2005 on ship-source pollution and on the introduction of penalties for infringements [2005] OJ L255/11 and Directive 2005/33/EC of the European Parliament and of the Council of 6 July 2005 amending Directive 1999/32/EC as regards the sulphur content of marine fuels [2005] OJ L191/59.

[34] International Convention for the Prevention of Pollution from Ships of 2 November 1973, 1340 UNTS 184; Protocol of 17 February 1978, 1340 UNTS 61.

[35] See Henrik Ringbom, *The EU Maritime Safety Policy and International Law* (Nijhoff 2008) 48 et seq.

[36] See n. 33.

[37] MARPOL Annexes I and II are automatically binding for all parties to the treaty in accordance with Art. 14 MARPOL. They address marine pollution by oil and dangerous liquid substances carried in bulk.

on the Law of the Sea (UNCLOS).[38,39] The High Court of Justice stayed the proceedings and submitted the question at issue to the ECJ for a preliminary ruling in accordance with Art. 267 TFEU.

In its judgement of 3 June 2008, the Court refused to measure the Directive's provisions against the requirements of international law, whose breach was being invoked. Whether the claims raised by the plaintiffs in the initial proceedings were founded was not ruled upon. With regard to MARPOL, the ECJ relied on the formal argument that the EU was not party to the treaty[40] and—unlike in the case of GATT 1947[41]—also had not assumed the competences of the Member States in that respect.[42] In addition, the Court argued that the relevant provisions contained in the MARPOL Annexes would not represent binding customary international law.[43] In respect of UNCLOS, the ECJ held that, although the Convention would be binding for Union organs in accordance with Art. 216 (2) TFEU (= Art. 300 (7) EC), it '[...] does not establish rules intended to apply directly and immediately to individuals and to confer upon them rights or freedoms capable of being relied upon against States, irrespective of the attitude of the ship's flag State.'[44] According to the Court, it would only be able to judge the validity of a rule in light of an international treaty if the nature and substance of that treaty would not stand in opposition and if its provisions would be sufficiently precise and unconditional in form and content.[45]

3.2. *Critical Appraisal of the Decision*

3.2.1. *On the Direct Applicability and Executability of Older National Treaties*

It is submitted that the ECJ's reference to the lack of direct legal effect on individuals of UNCLOS rules is, in principle, not convincing. The EU acceded to UNCLOS in 1998[46] and has both exclusive and shared

---

38 United Nations Convention on the Law of the Sea of 10 December 1982, 1833 UNTS 397.

39 ECJ, Case C-308/06 *Intertanko and Others* [2008] ECR I-4057, para 37 et seq.

40 Ibid., para 47; cf. already ECJ, Case C-379/92 *Peralta* [1994] ECR I-3453, para 16.

41 ECJ, Cases C-21–24/72 *International Fruit Company*, [1972] ECR 1219, para 10 et seq.

42 ECJ, Case C-308/06 *Intertanko and Others* [2008] ECR I-4057, para 48 et seq.

43 Ibid., para 51.

44 Ibid., para 64.

45 Ibid., para 45.

46 Cf. Council Decision 98/392/EC of 23 March 1998 concerning the conclusion by the European Community of the United Nations Convention of 10 December 1982 on the Law

competences in regard to particular areas codified in the treaty.[47] Precondition for the review of an act of secondary EU law as part of a preliminary ruling and in light of a 'communitized' international treaty in terms of Art. 216 (2) TFEU is categorically not the direct applicability of the norms contained in the treaty—that is, the possibility of the direct invocation of the pertinent norms by individuals in the sense of the *Van Gend en Loos* judgement.[48] On the contrary, it is only necessary that the pertinent norms are self-executing—that is, that they have direct effect.[49] In most instances, the ECJ did not rely on the requirement of direct applicability that it used in the *Intertanko* judgement[50] when assessing its competence to review an act of secondary EU law, but instead, while referring to the conceptually confusing term 'direct applicability', in substance analyzed the executability of the relevant treaty norms.[51] In the words of Advocate General *Kokott*:

> Whether or not the provisions in question are directly applicable and—correlatively—whether they confer rights on individuals is not conclusive for the purpose of responding to the request for a preliminary ruling. Even the legal bases of the Treaties are in principle not directly applicable in the sense that individuals can derive from them rights or legal consequences to their benefit. Nevertheless, individuals may question the legality of rules of secondary law by contesting the legal basis thereof.[52]

---

of the Sea and the Agreement of 28 July 1994 relating to the implementation of Part XI thereof [1998] OJ L179/1.

[47] Cf. Declaration concerning the competence of the European Community with regard to matters governed by the United Nations Convention on the Law of the Sea of 10 December 1982 made pursuant to article 5 (1) of Annex IX of the Convention [1998] OJ L179/129.

[48] Cf. ECJ, Case 26/62 *van Gend en Loos* [1963] ECR 3, 24 et seq.

[49] See already Proelss (n. 16) 427 et seq.; ibid., 'European Community Law and WTO Regulations: The Direct Effect-Doctrine Revisited', in Suthiphand Chirathivat, Franz Knipping, Cillian Ryan and Paul. J.J. Welfens (eds), *EU—ASEAN. Facing Economic Globalisation* (Springer 2009) 193, 195 et seq.; Mario Mendez, 'The Legal Effect of Community Agreements: Maximalist Treaty Enforcement and Judicial Avoidance Techniques' (2010) 21 EJIL 83, 102.—Precisely this executability is meant when the ECJ stated in the *Intertanko* judgement that the 'nature and the broad logic' of UNCLOS may not stand in the way of a review of secondary law in its light (ECJ, Case C-308/06 *Intertanko and Others* [2008] ECR I-4057, para 45). Cf. however ECJ, Case C-366/10 *The Air Transport Association of America and Others*, Opinion of Advocate General Kokott, para 71 et seq. (for legal proceedings brought by individuals); in contrast, the ECJ did not address whether the rights of individuals had been directly affected in its judgement of 21 December 2011; cf. para 53 of the judgement.

[50] ECJ, Case C-308/06 *Intertanko and Others* [2008] ECR I-4057, para 64.

[51] Cf. for example ECJ, Case C-104/81 *Kupferberg* [1982] ECR 3641, para 22 et seq.; Case C-12/86, *Demirel* [1987] ECR 3719, para 14; Case C-344/04 *IATA* [2006] ECR I-403 para 39; Case C-366/10 *The Air Transport Association of America and Others*, para 54 et seq.

[52] ECJ, Case C-308/06 *Intertanko and Others* [2008] ECR I-4057, Opinion of Advocate General Kokott, para 66.

It is not at the Court's discretion to establish additional criteria beyond those provided by the TFEU for reviewing an act of secondary law in light of a treaty already adopted by the EU.[53] In this regard the judgement in the *Intertanko* case has been justifiably criticized for relying on 'judicial avoidance techniques';[54] it can in fact be assumed that many UNCLOS norms are self-executing—at least to the extent that they indirectly reference MARPOL by using the term 'generally accepted international rules and standards' (cf. for example Art. 21 (2), Art. 211 (2) and (5) UNCLOS).[55]

Upon closer examination, however, it should be taken into consideration that a special situation lay at the root of the *Intertanko* judgement.[56] The proceedings before the High Court of Justice were initiated by the plaintiffs with the sole purpose of bringing about a direct review of Directive 2005/35/EC in light of international law. At the time the proceedings were initiated the Directive had not yet been transposed into national law in the United Kingdom. In this regard, there was a danger that the comparatively high barriers to individual actions before the ECJ would in this way be circumvented. It is well known that natural and legal persons may only 'institute proceedings against an act addressed to that person or which is of direct and individual concern to them, and against a regulatory act which is of direct concern to them and does not entail implementing measures' (Art. 263 (4) TFEU). According to the so-called Plaumann formula, direct and individual concern can be assumed when the act being challenged 'affects [the plaintiff] by reason of certain attributes which are peculiar to them or by reason of circumstances in which they are differentiated from all other persons, and by virtue of these factors distinguishes them individually just as in the case of the person addressed.'[57] It is submitted that in the case at hand, the ECJ's insistence on the direct

[53] See however ibid., para 67.

[54] Mendez (n. 49) 99 et seq.

[55] Similar ECJ, Case C-308/06 *Intertanko and Others* [2008] ECR I-4057, Opinion of Advocate General Kokott, para 46 et seq.

[56] See Sonja Boelaert-Suominen, 'The European Community, the European Court of Justice and the Law of the Sea' (2008) 23 IJMCL 643, 702, 708.

[57] ECJ, Case C-25/62 *Plaumann* [1963] ECR 95, headnote 4.—Whether the reformulation of the basis for legal action in Art. 263 (4) Alt. 2 TFEU has retroactive effects on the interpretation of individual concern and prompts a departure from the *Plaumann* formula is not uniformly judged; see Matthias Kottmann, 'Plaumanns Ende: Ein Vorschlag zu Art. 263 (4) AEUV' (2010) 70 ZaöRV 547, 556 et seq.

applicability of UNCLOS norms served indirectly to uphold the requirements of Art. 263 (4) TFEU.[58]

Having said that, such a synchronizing of the requirements of European judicial protection for individuals in the context of Art. 263 TFEU and Art. 267 TFEU seems to be inconsistent with the requirements of the rule of law if viewed in light of the principle of effective legal protection (cf. Art. 2 TEU).[59] Early on, the ECJ itself recognized the need to find a balance between the law of effective legal protection, on the one hand, and the imperative it set for itself to restrictively interpret the admissibility requirements for individual claims, on the other. According to the Court, this need arises from the nature of the EU as a community of law, in which neither the Member States nor the organs of the Union may evade review of whether their actions are in accordance with the founding treaties.[60] In order to ensure a 'complete system of legal remedies' in situations where affected parties are barred from bringing direct actions, 'such persons may plead the invalidity of general measures before the national courts and cause the latter to request the Court of Justice for a preliminary ruling.'[61] This countervailing mechanism fails, however, when—as in the case of the *Intertanko* judgement—the requirements for an individual action are nonetheless backhandedly applied during preliminary proceedings. It is precisely that moment when the character of the preliminary ruling process fails to foster the legal protection for individuals originally demanded by the ECJ itself. This seems even more problematic in light of Art. 19 (1) 2 TEU, according to which 'Member States shall provide remedies sufficient to ensure effective legal protection in the fields covered by Union law.' While this rule, which was included in the TEU with the Treaty of Lisbon, is not directly applicable to the institutions of the EU, it is, arguably, nonetheless an expression of the need for comprehensive legal protection. Its relevance becomes evident if one takes into account that the Member States' courts are generally not entitled to overrule EU legislative acts due to the fact that the Member States have transferred the respective competence to the ECJ by way of Art. 19 (3) (b) TEU and Art. 263 TFEU. Thus

---

58 Cf. Eileen Denza, 'A Note on Intertanko' (2008) 33 ELR 870, 875 as well as ECJ, Case C-366/10 *The Air Transport Association of America and Others*, Opinion of Advocate General Kokott, para 66.

59 Consolidated Version: [2010] OJ C 83/13.

60 ECJ, Case C-294/83 *Les Verts* [1986] ECR 1339, para 23.

61 Ibid.; see also Meinhard Schröder, 'Die Vorlagepflicht zum EuGH aus europarechtlicher und nationaler Perspektive' [2011] EuR 808, 810.

the criticism ultimately remains that in *Intertanko* the ECJ unjustifiably evaded making a decision on the matter.[62]

3.2.2. *On the Interpretation of European Union Law in a Friendly Manner towards International Law*

It is submitted that the same can be said concerning MARPOL. In this regard, the judgement is arguably based on an exaggerated and one-sided understanding of the autonomy of the EU legal order. It is true that the EU is not party to the Convention; it also has not assumed the substantive rights and duties of the Member States by way of substitution. Nonetheless, the Court failed to take the collision clause contained in Art. 351 (1) TFEU sufficiently into account.[63] With regard to the object and purpose of that clause and in accordance with the maxim *pacta sunt servanda*, EU law shall not undermine the rights and obligations of the Member States established by international treaties.[64] According to the majority view, this norm, which is literally only applicable to treaties adopted prior to 1 January 1958 ('older national treaties'), can be applied by way of analogy to treaties adopted after 1 January 1958 if and to the extent that these collide with primary or secondary law which came into existence after their adoption.[65] At the same time, para 2 of the norm obliges the Member States to utilize all appropriate means to resolve incompatibilities between the international treaties they have adopted and EU law. This may necessitate negotiated revisions to the treaties[66] or, ultimately, even

---

[62] Yet another question would be the result that the ECJ needed to reach in the matter. That cannot be answered here. See for example *Ringbom* (n. 35) 401 et seq.

[63] Reaching too far is ECJ, Case C-308/06 *Intertanko and Others* [2008] ECR I-4057, Opinion of Advocate General Kokott, para 77.

[64] In general Paul Craig and Gráinne de Búrca, *EU Law: text, cases and materials* (5th ed, Oxford Univ Press 2011) 204; Kirsten Schmalenbach, in Christian Calliess and Matthias Ruffert (eds), *EUV/AEUV*, (4th ed, Beck 2011) Art. 351 mn. 1.

[65] Cf. Craig and de Búrca (n. 64) 204; Stefan Lorenzmeier, in Eberhard Grabitz, Meinhard Hilf and Martin Nettesheim (eds) *Das Recht der Europäischen Union* (Vol. III, Status: 46. Ergänzungslieferung Beck 2011) Art. 351 mn. 24; Schmalenbach (n. 64), mn. 8; André Nollkaemper and Ellen Hey, 'Implementation of the LOS Convention at Regional Level: European Community Competence in Regulating Safety and Environment Aspects of Shipping' (1995) 10 IJMCL 281, 298; Proelss (n. 16), 334. See also ECJ, Case C-188/07 *Commune de Mesquer* [2008] ECR I-4501, Opinion of Advocate General Kokott, para 95: 'A mutatis mutandis application of the first paragraph of Article 307 EC does not lead to any other conclusion. It is conceivable where an international obligation on the part of a Member State conflicts with a subsequently agreed measure of secondary law.'—If the Member States adopt a treaty which violates existing secondary law, this constitutes a violation of the principle of cooperation in accordance with Art. 4 (3) TEU; see Lorenzmeier, l.c.

[66] Cf. only ECJ, Opinion 1/76 [1977] ECR 741, para 7.

their suspension or termination.[67] This obligation exists only to the extent that the treaties are incompatible with EU law, however.[68] Whether or not this is the case must be determined on an individual basis. The ECJ in *Intertanko* refused to perform this test although it had already found in its previous case law that Art. 351 (1) TFEU would fail to serve its purpose 'if it did not imply a duty on the part of the institutions of the Community not to impede the performance of the obligations of Member States which stem from a prior agreement.'[69]

It is not necessary here to reassess in detail a line of argument advocated mainly by German authors which argued in favor of a limitation of EU competences stemming from the aforementioned obligation, provided that all Member States have acceded to the treaty in question.[70] Despite the fact that this view was well-received in legal literature[71] and, indeed, could have provided support for the plaintiffs' argumentation in the initial proceedings leading up to the *Intertanko* judgement, it is arguably going too far in light of Art. 351 (2) TFEU to posit that a breach of para 1 of that provision implies a violation of competences on behalf of the EU organs, thus rendering the legal act in question invalid (and not merely unlawful due to the breach of Art. 351 (1) TFEU). The nature of Art. 351 TFEU, which requires that the potential for collisions exists as a matter of logic, is a further argument against a competence-limiting effect.[72] Nonetheless,

---

67 For example ECJ, Case C-13/93 *Minne* [1994] ECR I-37, para 15.

68 On violations of Art. 351 (2) TFEU see ECJ, Case C-205/06 *Commission/Austria* [2009] ECR I-1301, para 38 et seq. and Case C-249/06 *Commission/Sweden* [2009] ECR I-1335, para 39 et seq.

69 ECJ, Case 812/79 *Burgoa* [1980] ECR 2789, para 9; cf. also ECJ, Case C-84/98 *Commission/Portugal* [2000] ECR I-5215, para 53; Case C-466/98 *Commission/United Kingdom* [2002] ECR I-9427, para 23 et seq.; Case C-216/01 *Budějovický Budvar* [2003] ECR I-13617, para 145. In Case C-158/91 the ECJ examined whether a Community rule can be 'derived of effect by an earlier international agreement' ([1993], ECR I-4300, para 13).

70 Proelss (n. 16) 330 et seq.; Wolfgang Graf Vitzthum, 'Europäisches Seerecht. Eine kompetenz rechtliche Skizze' in Michael Brenner, Peter M. Huber and Markus Möstl (eds), *Der Staat des Grundgesetzes—Kontinuität und Wandel: Festschrift für Peter Badura zum siebzigsten Geburtstag* (Mohr-Siebeck 2004) 1189, 1204 et seq.; Hans Krück, *Völkerrechtliche Verträge im Recht der Europäischen Gemeinschaften* (Springer 1977) 139 et seq.; also Nollkaemper and Hey (n. 65) 298.

71 See Doris König, 'The EU Directive on Ship-Source Pollution and on the Introduction of Penalties for Infringements: Development or Breach of International Law?' in Tafsir Malick Ndiaye and Rüdiger Wolfrum (eds), *Law of the Sea, Environmental Law and Settlement of Disputes, Liber Amicorum Thomas Mensah* (Martinus Nijhoff 2007) 767, 784; Henning Schult, *Das völkerrechtliche Schiffssicherheitsregime* (Duncker und Humblot 2005) 332 et seq.

72 Lorenzmeier (n. 65), para 18. Against this background, I no longer maintain my earlier opinion to the contrary (cf. Proelss [n. 16] 332).

it should clearly be noted that a danger exists in situations where the Member States are bound to maximum standards at the international level[73] that the autonomous legislative activities of the Union could bring them into an irreconcilable web of obligations as a consequence. Having said this, on closer observation one can only speak of a collision in a narrower sense if and to the extent that the treaty in terms of Art. 351 (1) TFEU establishes more than mere minimum standards, which can be transgressed either unilaterally or at the supranational level. This cannot generally be said in regard to MARPOL.[74]

Irrespective of the aforementioned, it is submitted that the object and purpose of Art. 351 (1) TFEU and the EU's commitment to the compatibility of its legal system with international law that is implicit in that provision as well as in Art. 3 (5) TEU[75] in conjunction with the principle of sincere cooperation and mutual respect (cf. Art. 4 (3) TEU), support the premise that EU law must at least be interpreted in a manner friendly toward the international rights and obligations of the Member States deriving from treaties in terms of Art. 351 (1) TFEU. In light of the principle of uniform application of EU law as well as the wording of Art. 351 TFEU, this premise should also apply where only some (and not all) Member States are contractually bound.[76] Following case-law of the German Federal Constitutional Court (*Bundesverfassungsgericht*) concerning the interpretation of national law with affinity toward international law,[77] the rule of interpretation advocated here implements the maxim *pacta sunt servanda* expressed in Art. 351 (1) TFEU in situations in which the rules contained in older national treaties do not appear to be compatible with EU law at first glance. From the standpoint of legal methodology, it constitutes a conflict prevention tool, which can only be applied if and to the extent to

73 Maximum standards are standards which may not be exceeded by the contracting parties.

74 See Schult (n. 71) 336 et seq.; too undifferentiated is Sara Vatankhah, 'The Contribution of the EC to Maritime Safety in View of the "Third Maritime Safety Package" of the European Commission' in Peter Ehlers and Rainer Lagoni (eds), *Maritime Policy of the European Union and Law of the Sea* (LIT 2008) 41, 57. Cf. also ECJ, Case C-308/06 *Intertanko and Others* [2008] ECR I-4057, Opinion of Advocate General Kokott, para 80 et seq.

75 According to Art. 3 (5) TEU the European Union shall 'contribute to peace, security, the sustainable development of the Earth, solidarity and mutual respect among peoples, free and fair trade, eradication of poverty and the protection of human rights, in particular the rights of the child, as well as to the strict observance and the development of international law, including respect for the principles of the United Nations Charter.'

76 Proelss (n. 16) 333; potentially of a different opinion Schult (n. 71) 334.

77 Cf. only BVerfGE 58, 1 (34); 59, 63 (89); 64, 1 (20).

which a norm of EU secondary law allows several interpretations. In any event, existence of a duty to interpret EU law with affinity toward international law ought to be considered in situations in which the Member States had shared external competence in regard to the subject matter governed by the older national treaty at the time that treaty was concluded.[78] A further condition ought to be that the Member States do not enter into treaty obligations with the purpose of circumventing EU law. The latter results from the principle of sincere cooperation as well as from the ratio of para 2 of Art. 351 TFEU. These provisions also lead to the result that in situations where an interpretation of EU law with affinity toward international law fails due to the conflicting wording of the pertinent norm, the obligation of the EU to not undermine the rights and obligations contained in the pertinent older national treaty can only exist for a brief transitional period, which is difficult to tangibly define.

By preventing such situations whenever possible, the duty to interpret EU law in a manner friendly toward international law advocated here—which in light of Art. 3 (5) TEU must also be applied to customary international law[79]—helps to strike a balance between the prohibition of interference contained in Art. 351 (1) TFEU and the duty to adapt contained in Art. 351 (2) TFEU, thus contributing to legal certainty. Moreover, it guarantees that the Member States will not face the danger of being held responsible for violations of international law, which have arisen due to actions in compliance with EU law.[80] Interestingly, the ECJ expressed itself in a similar way in the *Intertanko* judgement when it found that Member States' obligations under MARPOL can influence the interpretation of UNCLOS and provisions of EU secondary law falling within the scope of MARPOL:

---

[78] Cf. Proelss (n. 16) 335 et seq.—If a Member State adopts an international treaty on a subject matter over which the Union has exclusive competence, this violation of EU competences does not have any influence on the validity of the treaty under international law. Within the EU, however, Art. 351 (1) is not applicable to such situations. Cf. Lorenzmeier (n. 65), para 19.

[79] The ECJ has consistently assumed an obligation on the part of Union organs to observe customary international law in their exercise of competences; cf. ECJ, Case C-286/90 *Poulsen und Diva Navigation* [1992] ECR I-6019, para 9; Case C-158/91 *Levy* [1993] ECR I-4300, para 13; Case C-405/92 *Mondiet* [1993] ECR I-6133, para 13; Case C-364 and 365/95 *T. Port* [1998] ECR I-1023, para 60 et seq.; Case C-162/96 *Racke* [1998] ECR I-3655, para 45; Case C-366/10 *The Air Transport Association of America and Others*, para 123.

[80] Cf. BVerfGE 74, 358 (370).

> In view of the customary principle of good faith, which forms part of general international law, and of Article 10 EC, it is incumbent upon the Court to interpret those provisions taking account of Marpol 73/78.[81]

Due to the nature of the national Court's reference, however, which was restricted to the examination of the validity of the specific Directive, the Court did not undertake an interpretation of EU law in the manner required of it. This reticence reflects the far too formalist tone of the entire judgement and leads to a less than satisfying result in the matter. For the referring court, *Intertanko* only concerned the compatibility of Directive 2005/35/EC with the MARPOL Convention. If the ECJ explicitly assumed that the requirements of MARPOL ought to be upheld when interpreting the provisions of Directive 2005/35/EC precisely because all Member States (although not the EU itself) are bound by the Convention, it arguably would have needed to clarify whether such an interpretation of the Directive is at all possible in regard to the question submitted for the preliminary ruling and, if so, which consequences then emerge.

There are good arguments for applying the principle of interpretation of EU law in a manner friendly toward international law to other problematic areas in the external relations of the EU and its Member States. Emerging issues such as those arising in the context of overlapping jurisdictions between courts and arbitral tribunals could be resolved in this manner. An important example is the *MOX Plant* case, in which Ireland submitted claims against the United Kingdom before both an arbitral tribunal in accordance with Annex VII UNCLOS as well as before an arbitral tribunal on the basis of Art. 32 of the Convention for the Protection of the Marine Environment of the North-East Atlantic (OSPAR Convention)[82] for violations of the relevant convention.[83] In addition to this, Ireland submitted a

---

[81] ECJ, Case C-308/06 *Intertanko and Others* [2008] ECR I-4057, para 52. In its recent judgement on emissions trading the ECJ did not consider an interpretation with affinity toward the Chicago Convention (cf. Case C-366/10, para 72). Cf. however the Opinion of Advocate General Kokott, para 66: 'The fact that all Member States of the European Union are Contracting Parties to the Chicago Convention can nevertheless have an effect on the interpretation of provisions of EU law; this follows from the general principle of good faith, which also applies under international law and has found specific expression under EU law in Article 4 (3) TEU' (footnotes omitted).

[82] Convention for the Protection of the Marine Environment of the North-East Atlantic of 22 September 1992, 2354 UNTS 67.

[83] Cf. *Dispute Concerning Access to Information under Article 9 of the OSPAR Convention (Ireland v. United Kingdom)*, Final Award of 2 July 2003, (2003) 42 ILM 1118 et seq. In contrast, the arbitral tribunal stayed the proceedings in accordance with Annex VII UNCLOS after the Commission made its intention public to institute infringement proceedings against Ireland before the ECJ; cf. *The MOX Plant Case (Ireland v. United Kingdom)*, Order

request for provisional measures in accordance with Art. 290 (5) UNCLOS to the International Tribunal for the Law of the Sea (ITLOS).[84] Finally, the EU Commission initiated infringement proceedings against Ireland before the ECJ on the basis of the existing obligation of Member States 'not to submit a dispute concerning the interpretation or application of the Treaties to any method of settlement other than those provided for therein' according to ex Art. 292 EC (= Art. 344 TFEU), and the premise that the treaties adopted by the EU and the Member States as mixed agreements represent integral parts of the legal order of the Union.[85] In its judgement, the Court followed the Commission's argumentation on the whole and stated the following:

> It follows that the provisions of the Convention relied on by Ireland in the dispute relating to the MOX plant and submitted to the Arbitral Tribunal are rules which form part of the Community legal order. The Court therefore has jurisdiction to deal with disputes relating to the interpretation and application of those provisions and to assess a Member State's compliance with them. [...] It is, however, necessary to determine whether this jurisdiction of the Court is exclusive, such as to preclude a dispute like that relating to the MOX plant being brought by a Member State before an arbitral tribunal established pursuant to Annex VII to the Convention. [...] The Court has already pointed out that an international agreement cannot affect the allocation of responsibilities defined in the Treaties and, consequently, the autonomy of the Community legal system, compliance with which the Court ensures under Article 220 EC. That exclusive jurisdiction of the Court is confirmed by Article 292 EC, by which Member States undertake not to submit a dispute concerning the interpretation or application of the EC Treaty to any method of settlement other than those provided for therein [...].[86]

The ECJ's assumption of exclusive jurisdiction for disputes over the rights and obligations of Member States arising from international treaties to which the EU is also party is again, arguably, based on an oversubscription to the autonomy of the Union's legal order. While it is true that Art. 216 (2) TFEU (and thus the ECJ's jurisdiction in principle, as well) extends to the provisions of a mixed agreement falling within the Union's

of 24 June 2003, (2003) 42 ILM 1187 et seq. Following the ECJ's judgement of 30 May 2006, Ireland withdrew its claim from the UNCLOS tribunal on 15 February 2007.

[84] Cf. *The MOX Plant Case* (*Ireland v. United Kingdom*), Order of 3 December 2001, ITLOS Reports 2001, 95 et seq.

[85] Cf. ECJ, Case C-459/03, *Commission/Ireland* [2006] ECR I-4635, para 61 et seq.

[86] Ibid., para 121 et seq.

sphere of competences,[87] this does not force one to conclude that calling on an international court or arbitral tribunal intended by the applicable treaty automatically constitutes a violation of Art. 344 TFEU—even if and to the extent that the Union has already adopted rules in the area in question. Indeed, instisting on the ECJ's exclusive jurisdiction fails to recognize that the adoption of a mixed agreement in no way affects the allocation of competences between the EU and the Member States in the internal legal order of the Union. Consequently, recourse by the Union organs to a shared competence on the field of external relations does not lead to an exclusive Union competence in terms of Art. 2 (2) sentence 2 TFEU. If the interpretation of the provisions of a mixed agreement is at issue which belong to an area that does not fall under the exclusive competence of the EU, it is not comprehensible in general why Art. 344 TFEU should then lead to the exclusive jurisdiction of the ECJ.[88] The mere acceptance of the jurisdiction of an international court or tribunal in regard to a mixed agreement does not, in itself, immediately endanger the division and exercise of the competences as laid out in the European treaties. Only when the court or tribunal's interpretation of the pertinent norms of the mixed agreement is no longer compatible with other Union law can this lead to the inapplicability of the relevant judgement within the EU legal order. The line of argument advocated here is consistent with the ratio of Art. 351 (2) TFEU (which is, however, not directly applicable to the constellation described above in light of the EU's status as contracting party), and it should be noted that the ECJ itself did not object to the EU's own acceptance of a dispute settlement mechanism governed by international law, provided that it did not compromise the foundations of the EU legal order.[89]

---

[87] In light of Art. 4 (3) TEU and the principle of uniform application of EU law it is submitted that this also applies to those provisions which fall under the exclusive competences of the Member States; cf. Alexander Proelss, 'The Intra-Community Effects of Mixed Agreements' in Paul J.J. Welfens, Franz Knipping, Suthiph Chirathivat and Cillian Ryan (eds), *Integration in Asia and Europe* (Springer 2006) 255 et seq. with references to ECJ case law. See also Pierre Pescatore, 'Les relations extérieures des communautés européennes', (1961–II) 103 RdC 1, 133.

[88] But see Boelaert-Suominen (n. 56) 678 et seq.—If one accepts the view presented here that the duty to interpret Union law with affinity toward international law is also applicable to courts and tribunals whose jurisdictions are established by treaties concluded as mixed agreements, a revision of Art. 344 TFEU is not necessary; cf. however Nikolaos Lavranos, 'The *MOX Plant* and *Ijzeren Rijn* Disputes: Which Court is the Supreme Arbiter?' (2006) 19 LJIL 223, 246.

[89] ECJ, Opinion 1/91 [1991] ECR I-6079, para 35.

It shall not be left unmentioned that the ECJ's approach to the autonomy of the Union legal order ultimately fosters disregard for the expertise of a court or tribunal in regard to its founding international treaty. Arguably, the court specifically created for the settlement of disputes under an international treaty is generally the best able to appropriately unfold the significance of the rights and obligations contained in that treaty in regard to its objectives and in consideration of its genesis.[90] Consideration of the interpretation of a mixed agreement by a specialized international body thus represents an ideal approach to the implementation of the objectives contained in Art. 3 (5) TEU. This is particularly true in respect of functional institutions (e.g. implementation committees) that were established in order to guarantee the proper performance of an international treaty, and thereby seek to prevent the emergence of legal disputes. In any case, whether or not Art. 344 TFEU is at all applicable to such non-judicial institutions seems questionable from the outset.

On the other hand, the application of the duty to interpret EU law with affinity toward international law in an institutional context does not stand in the way of Member States calling on the ECJ as an alternative; this principle of interpretation cannot alter the comprehensive jurisdiction of the ECJ over EU law (including international treaties which represent integral parts of the EU legal order in terms of Art. 216 (2) TFEU). What is argued here is that Art. 344 TFEU simply ought not be applied in a manner that creates an absolute barrier against the jurisdiction of international courts and tribunals within the EU legal order.

4. The Compatibility of the 'Erika III' Measures with International Law

In light of the aforementioned the question remains to be answered whether the 'Erika III' package is compatible with international law.

---

[90] Cf. also *The MOX Plant Case* (*Ireland v. United Kingdom*), Order of 3 December 2001, ITLOS Reports 2001, 95 (para 51). Also critical in regard to European jurisdiction in this area of environmental protection is Simon Marsden, 'MOX Plant and the Espoo Convention: Can Member State Disputes Concerning Mixed Environmental Agreements be Resolved Outside EC Law?' (2009) 18 RECIEL 312, 326: '[...] may not produce the best outcome [...]'.

4.1. *Insurance of Ship Owners for Maritime Claims*

Initially, doubts were raised as to the compatibility of Directive 2009/20/EC with the international law of the sea. The draft of this Directive originally provided that foreign ships must possess and demonstrate proof of insurance coverage as soon as they enter the Exclusive Economic Zone (EEZ) of an EU Member State. As the EEZ does not belong to a State's territory, this could have represented an exercise of extraterritorial jurisdiction[91] not covered by the UNCLOS (cf. Art. 58 (1)) and the 1976 Convention on Limitation of Liablity for Maritime Claims.[92] Although a coastal State may adopt laws and regulations for the prevention, reduction and control of marine pollution from ships in its EEZ in accordance with Art. 211 (5) UNCLOS, the rules in question must comply with the 'generally accepted international rules and standards' (GAIRAS) adopted under the auspices of the IMO.[93] Such standards did not and do not exist in regard to liability for maritime claims.

However, these concerns were addressed in the final version of the Directive. Its Art. 4 (2) limits the scope of application to ships calling at the ports of EU Member States. As part of the coastal State's internal waters, ports are subject to its full territorial sovereignty.[94] The extent to which Member States are additionally authorized to demand compliance with this requirement on ships travelling in their territorial waters—this formulation refers to the territorial sea falling under the 'aquitorial sovereignty' of the coastal State, in which foreign ships enjoy the right of innocent passage (cf. Art. 18 et seq. UNCLOS)— is conditional on its compatibility with international law. As the Directive fully reflects the requirements of the 1976 Convention on the Limitation of Liability for Maritime Claims in regard to the maximum amount of liability (cf. Art. 4 (3)), it is clearly consistent with the requirements of international law.

---

[91] Vatankhah (n. 74) 62 et seq.; Jenisch (n. 17) 56.

[92] See n. 32.

[93] According to Eric Molenaar (*Coastal State Jurisdiction over Vessel-Source Pollution* [Springer Netherland 1998] 175 et seq.), it is not necessary for the pertinent rules and norms to be customary international law in order for them to be considered 'generally recognized'; the only prerequisite for this is that the relevant IMO rules have entered into force. Cf. also Ringbom (n. 35) 433.

[94] Cf. only Robin R. Churchill/Vaughan A. Lowe, *Law of the Sea* (3rd edn, JurisPubl 1999) 64 et seq.; Ringbom (n. 35) 204 et seq. with further references.

4.2. *Notification Requirement in Accordance with the Directive on Port State Control*

This could be a different matter in regard to Directive 2009/16 on port State control, however. Problematic at first glance is the duty arising from Art. 9 in conjunction with Annex III of the Directive to notify the port authority at least three days prior to the expected arrival of a ship in its harbor for the purpose of vessel identification, expected length of anchorage, and so forth. It often cannot be ruled out that a ship is still located outside the maritime aquitory of the coastal State in question at that point in time. In light of the freedom of navigation applicable beyond the territorial sea (cf. Art. 58 (1) in conjunction with Art. 87 (1)(a) UNCLOS) there is good reason to argue that a third State vessel cannot be required to notify of its arrival when it is still in the EEZ or on the high seas.[95] Admittedly, the ECJ has refused to regard the inclusion of non-European aircraft operators in the European trading scheme for greenhouse gas emissions allowances as an impermissible exercise of extraterritorial jurisdiction,[96] even though the emissions certificates to be traded must be calculated for the entire international flight (and not just for the individual segments over the territories of the EU Member States). This does not lead to a different outcome, though. Schematic recourse to the argument that the flights in question start or end at an airport on EU territory[97] does not change the fact that the principle of territoriality fails to provide a sufficient link for the inclusion of flight segments over non-EU territory in the trading scheme. The fact that a State has jurisdiction over a particular subject matter (here: take-off and landing on its own territory) does not necessarily imply that its legislative authority extends to all other aspects related to the subject matter concerned.[98] This is all that much more true as the possibility exists of restricting the inclusion of non-European aircraft operators in the European emissions trading scheme to flight

[95] See Rainer Lagoni, 'Vorsorge gegen Schiffsunfälle im Küstenvorfeld: Gemeinschaftliches Schiffsmeldesystem und Hafenzugang im Notfall' [2001] TranspR 284, 286; Vitzthum (n. 6) 176 et seq.; contra Ringbom (n. 35) 253 et seq.—In contrast, calling at a port itself can be made dependent on the fulfillment of the duty to notify; cf. Art. 25 (2) and Art. 211 (3) UNCLOS.

[96] ECJ, Case C-366/10 *The Air Transport Association of America and Others*, para 125 et seq.

[97] Ibid., para 125.

[98] See Markus Volz, *Extraterritoriale Terrorismusbekämpfung* (Duncker und Humblot 2007) 223.

segments over EU territory.[99] At any rate, comparability is lacking between the situation examined in the recent ECJ judgement and the situation at sea. This is because Art. 26 (1) UNCLOS establishes that even in the territorial sea no fees may be levied for mere passage (or for the 'use' of the environment of the coastal State) due to the right of innocent passage. For the maritime zones beyond the territorial sea this is all that much more applicable.

Notwithstanding this, the question whether a sufficient link for the duty to notify of arrival exists when a vessel is in the EEZ or on the high seas does not need a definite answer in the present context. As in the case of Directive 2002/59/EC on vessel traffic monitoring, the notification need not be made by the captain of the ship, but can alternatively be made by the agent or operator of a ship (cf. Art. 9 (1) of Directive 2009/16/EU). At least the shipping agent will regularly be present in the coastal State's territory. From this perspective the duty to notify does not represent an exercise of extraterritorial jurisdiction.[100]

4.3. *Access Refusals in Accordance with the Directive on Port State Control*

Finally, the question arises whether the possibility existing under the Directive on port State control of issuing temporary or even permanent refusals of access to ships of third States is compatible with the international law of the sea. The authorization to conduct port State controls in the first place derives from the fact that ports are subject to the coastal State's territorial sovereignty. Art. 211 (3) UNCLOS implies that port States have the competence to impose conditions on foreign vessels for access to their harbors or territorial waters or for anchoring at their off-shore terminals for the purpose of preventing, reducing and monitoring the pollution of the marine environment.[101] Art. 219 UNCLOS ('on their own initiative') further confirms the existence of a right on the part of the port State to conduct ship inspections. Almost all IMO Conventions contain provisions which authorize the treaty parties to inspect foreign ships calling at their ports in order to monitor compliance with the standards codified in the conventions.

---

[99] To the extent that the subject matter is divisible; cf. ibid., 223 with further references.

[100] On the Directive on vessel traffic monitoring see Proelss (n. 16) 402.

[101] Cf. also *Military and Paramilitary Activities in and against Nicaragua* (*Nicaragua v. USA*) (*Merits*), ICJ Reports 1986, 13, para 213.

The port State's authority to conduct inspections is not unlimited, however. Ships causing marine pollution on the high seas may, in general, only be pursued at ports of call in accordance with Art. 218 UNCLOS if and to the extent that violations of the international rules and norms applicable to discharge of materials or energy are at issue. This extraterritorial enforcement competence was extended to air pollution from shipping emissions with Regulation 11/6 of Annex VI MARPOL.[102]

If one looks beyond the question of the permissibility of access refusals, Directive 2009/16 does not exceed the boundaries set by UNCLOS or the standards set in the IMO Conventions. It does not lay down its own substantive requirements, but instead sets out exclusively procedural requirements for the EU-wide harmonization of standards for controls. In regard to the criteria for the detention of a third State vessel, Annex X of the Directive expressly refers to the rules contained in the pertinent IMO Conventions (including MARPOL) and thereby binds all control competences on the globally applicable maritime safety regime. In addition to this, Art. 4 (1) of the Directive requires, in general, that port State controls are performed 'in accordance with international law'. On the basis of the duty to interpret the norms of the Directive in conformity with international law following from that provision, the Member State competence set out in Art. 3 (1) to control a ship in its territorial waters but outside a port—somewhat problematic in regard to the right of innocent passage—still appears compatible with international law.

Can this also be taken to mean that a third State vessel can be permanently banned from European ports in the event of repeated violations of the substantive safety standards required under the IMO conventions? On the one hand, neither UNCLOS nor the IMO Conventions expressly provide for the imposition of a permanent refusal of access;[103] on the other hand, such a ban embodies a measure that falls within the territorial sovereignty of the port State as a matter of principle, and which is only prohibited when it is tied to the extraterritorial activities of a ship or its crew. The refusal of access must, however, be compatible with the

102 Admittedly, this only applies between States which have both ratified Annex VI MARPOL and are also parties to UNCLOS.—Regulation 11/6 states: 'The international law concerning the prevention, reduction and control of pollution of the marine environment from ships, including that law relating to the enforcement and safeguards, in force at the time of application or interpretation of the Annex, applies, *mutatis mutandis*, to the rules and standards set forth in this Annex' (emphasis in the original).

103 See Ringbom (n. 35) 295.

duty contained in Art. 227 UNCLOS not to discriminate against the ships of foreign States as well as the principle of proportionality—a general rule of law.[104] Thus, it would be incompatible with international law if an inspection practice were to emerge, according to which only ships flying particular flags would be denied access to European ports *a priori*—that is, without reasonable grounds for suspicion. However, Directive 2009/16 prevents violations of these principles by ensuring that (1) the imposition of a permanent refusal of access is only permissible after the third temporary refusal of access (cf. Art. 16 (4)), and that (2) legal action against the imposition of—temporary or permanent—refusals of access is possible (Art. 20). A violation of the principle of non-discrimination is not apparent because the Directive applies to all States, whose ships call at European ports. That the ships of one or more States could be affected by a refusal of access does not fall within the scope of application of Art. 227 UNCLOS; the imposition of control measures is, in reality, the consequence of the uniform application of standards applicable to all States.

In summary, the measures contained in the 'Erika III' package are compatible with the international law of the sea. A conflict between the EU, on the one hand, and the IMO, on the other, does not exist.

5. Conclusion

Although European secondary law is compatible with the requirements of international law in the given case, this cannot lead to the conclusion that a discussion of the extent of the Union's competences in the field of international relations—for example in regard to the extension of the European emissions trading scheme to non-European airlines—is much ado about nothing. That it could eventually come to a conflict between a globally responsible international organization, in which the EU does not directly participate, and the supranational EU is highlighted by previous experiences in the field of maritime safety. It should be recalled that the EU's 'Erika I' initiative to regulate the accelerated introduction of double hulls[105] motivated the IMO to resume its work on the topic and ultimately adopt an accelerated schedule for the decomissioning of single-hulled oil

[104] See also ibid., 296 et seq.
[105] Cf. Communication COM (2000) 142 final of 21 March 2000 from the Commission to the European Parliament and the Council on the safety of the seaborne oil trade.

tankers. While the EU adapted the original Commission draft in order to coordinate it with the new MARPOL requirements and ensure the compatibility of EU law with international law,[106] the readiness of the Union's organs, confronted with the oil spill on the coast of Brittany, to tread their own uniquely European path if need be was unmistakeable.[107] The EU evidently pursued a similar 'bottom-up approach' to the implementation of more stringent environmental standards in the area of maritime emissions.[108] In a communication of 28 January 2009, the Commission urged that '[i]f no effective global rules to reduce GHG emissions from [maritime transport] can be agreed upon', irrespective of the IMO's general responsibility for the drafting and adoption of global measures, the EU 'should agree its own measures.'[109] In another Communication it continued that the EU should:

> [...] strive for and cooperate towards achieving all the objectives of the EU maritime safety and security policies by means of international instruments agreed through the IMO. If IMO negotiations should fail, however, then the EU should take the lead in implementing measures on matters that are of particular importance for the EU, as a first step, pending wider international agreement and taking the international competitive environment into consideration.[110]

There is clearly no lack of will to undertake such a unilateral approach as seen in regard to the inclusion of foreign aviation in the EU-wide emissions trading scheme—after the ECJ's resolution of the outstanding legal questions in this context on 21 December 2011[111]—which became effective as of 1 January 2012.[112]

---

[106] Cf. Regulation (EC) No 417/2002 (n. 2).

[107] See already COM (2001) 370 final of 12 September 2001, European transport policy for 2010: time to decide, 97: 'The Community has built up a considerable body of law over the last ten years, especially in air and sea transport [...] The Community has adopted specific regulations which do not always coincide with the recommendations and agreements made in international organisations.'

[108] Overview in Alexander Proelss and Killian O'Brien, 'Völker- und europarechtliche Anforderungen an Abgasemissionen von Seeschiffen' [2011] NordÖR 97 et seq.

[109] COM (2009) 39 final of 28 January 2009, Towards a comprehensive climate change agreement in Copenhagen, 7 et seq.

[110] COM (2009) 8 final of 21 January 2009, Strategic goals and recommendations for the EU's maritime transport policy until 2018, 10.

[111] Cf. ECJ, C-366/10 *The Air Transport Association of America and Others*.

[112] Directive 2008/101/EC of the European Parliament and of the Council of 19 November 2008 amending Directive 2003/87/EC so as to include aviation activities in the scheme for greenhouse gas emission allowance trading within the Community [2009] OJ L8/3.

It is unlikely that the EU is prepared to act outside the globally applicable framework of international law on a permanent basis, however. At the forefront of the European approach is the effort—strengthened by Europe's significant economic power[113]—to exert political pressure on the Member States of the IMO to more quickly bring about a reform of international maritime and climate protection law. This political interest, however rational, should not distract from the legal dimension analyzed in this paper. Despite the fact that the EU's approach has led in the past to a tightening of the international maritime safety regime, this does not yet make it legitimate in and of itself. A purely outcome-oriented evaluation would fail to recognize that a repeatedly unilateral approach on the part of the EU could ultimately undermine the authority and competence of the IMO. In light of the global nature of shipping and its importance for the world economy, more stringent regional rules would be a small victory at a very high price.[114] Ultimately, this could even contribute to the further fragmentation of international law; legal uncertainty and an encroachment on international law's overall claim to normative validity could be the consequence.

Even from the EU's perspective, there are concerns about such European *Sonderwege*. On one hand, such an approach could (further) reduce the EU's political weight in international relations; in the words of *Gráinne de Búrca* this would represent 'a significant departure from the conventional presentation and widespread understanding of the EU as an actor maintaining a distinctive commitment to international law and institutions.'[115] The case law of the ECJ—which, in the view represented here, significantly overemphasizes the autonomy of the Union legal order despite the requirements of European primary law—would find its match in the course of action taken by other Union organs.[116] In regard to the specific example of shipping, the danger exists that outflagging in favor of non-European shipping registers would begin.

---

113 See Liu Nengye and Frank Maes, 'The European Union and the International Maritime Organization: EU's External Influence on the Prevention of Vessel-Source Pollution' (2010) 41 JMLC 581, 590.

114 Cf. also Rainer Lagoni, 'Umwelt und Schiffssicherheit im Völkerrecht und im Recht der Europäischen Gemeinschaften' (1994) 32 AVR 382, 395 et seq.; Proelss (n. 16) 326; uncritical, in contrast Nengye and Maes (n. 113) 588 et seq., 594.

115 Gráinne de Búrca, 'The European Court of Justice and the International Legal Order after *Kadi*' (2010) 51 HIJL 1, 2.

116 See also the criticism expressed in Mendez (n. 49) 103.

The European Union is therefore well advised to remember its role as a potential enforcement organ for the IMO. It has done so with the 'Erika III' package. In this area of international relations that concerns the effective implementation and enforcement of global standards, the supranational structure of the Union has stood the test—not least of which being due to the existence of compulsory jurisdiction. At the same time, however, some things still need improvement in regard to the centralized control of the relevant actors in the area of maritime safety—which is necessary for the effectiveness of any safety regime. Above all, the successful intensification of port State controls deserves unconditional support. It remains to be seen whether the Union organs will follow this path in the future, however—authentically fostering international law rather than persisting with an exaggerated understanding of the autonomy of the European Union's legal system.

Working Groups

Emissions Reduction and Emissions Trading Systems in Shipping

EMISSIONS REDUCTION AND EMISSIONS TRADING SYSTEMS IN SHIPPING: A BIMCO PERSPECTIVE

Torben Skaanild[1]

We all carry the responsibility to contribute to and perform a reasonable role with respect to reversing the negative effects of CO_2 emissions on the climate.

All industries, including shipping and aviation, should obviously also contribute—first and foremost by implementing energy efficient measures to reduce emissions.

Reasonable and realistic targets for the level of reduction should be set for each industry by governments in proportion to the amount of CO_2 emitted by that particular industry.

In order to assess what constitutes a realistic target, exact and reliable data must be obtained.

BIMCO supports transparent, predictable and enforceable regulatory instruments with global application that drives energy efficient solutions. Such regulations should be established by the IMO because it is the only organisation with sufficient technical capability on an international level to deal with regulating CO_2 emissions for shipping.

IMO recently adopted the Energy Efficiency Design Index (EEDI), which sets out the technical standards applicable from 2013 for certain categories of new ships irrespective of flag. It has been estimated that the predicted improvement in energy efficiency would result in a 25%–30% reduction in CO_2 emission by 2030.

IMO also adopted the mandatory Ship Energy Efficiency Management Plan for all ships which will further contribute to the sector's reduction of CO_2 emissions—but it may be necessary to further address existing ships. It should, however, be kept in mind that as the cost of fuel account for up to 80% of daily operating costs, ship-owners and operators will already have a strong incentive to operate their ships with utmost energy efficiency.

The main objective remains to limit the emission of CO_2 from ships whilst fulfilling the industry's important role in the global supply chain.

[1] Secretary General of BIMCO.

That shipping is the most environmentally friendly and cost-effective mode of transportation per ton/mile is uncontested, evidence being that around 90% of goods are transported by sea. Ships and ship-owners should in principle be in compliance once the emission reduction targets set by governments are met. Only in the event such targets cannot be met, one may have to apply a globally acceptable Market Based Measure (MBM) to provide ship-owners with an alternative to expensive retrofitting of ships.

Any form of a MBM introduced to shipping must fulfill the nine IMO principles (Effective; Binding and equally applicable; Cost-effective; Limit distortion; Not penalizing trade and growth; Goal-based; Promote R&D; Accommodating energy-efficient technology; Practical, transparent, fraud-free and easy to administer) and must resolve the conflict between the 'Common But Differentiated Responsibilities' (CBDR) principle of UNFCCC and the non-discriminatory principle of the IMO Convention. This is a very fundamental condition and one that BIMCO will strongly defend, in line with BIMCO's objectives of promoting fair business practices and defending free trade as well as open access to markets.

In the context of a MBM, it must be emphasized that the shipping industry should not be a funding mechanism for resolving the global warming issue. Shipping should be treated equitably to other transport modes. We strongly believe that the burden of regulations aimed at reducing GHG emissions from ships should be cost-neutral to ship-owners if and when the reduction target is reached. BIMCO supports regulations which will reward owners who invest in low-carbon technology. Shipping should not be a 'cash cow' in the context of generating funds to counter climate change; we firmly believe that any financial contribution should be no more than shipping's share of the total GHG emissions (2.7%).

We do not believe that an Emissions Trading Scheme (ETS) can be effectively applied to shipping, owing to this industry's global nature, its mobility and its diversity. Although the ETS model is claimed to have worked in some regional and national regulations concerning land-based industries, the lack of a global carbon market is a significant impediment to applying a global ETS to shipping. Moreover, the effects of an ETS based on full auctioning will severely impact world trade and unfairly penalise shipping, as emissions from ships would be priced below a set cap.

Shipping is a global industry with all its players competing for the same cargo on reasonable equitable terms and thus any form of exemption favouring particular flags would create unfair trading practices.

IMO has already taken considerable steps to further address emissions from shipping and the Expert Group specifically formed to review the presently proposed MBMs have made available its initial and thorough comparative analysis of the options. This work has to continue through further evaluations, elaboration and development to bring forward workable solutions that can be implemented and effectively controlled by governments in a form that does not distort trade and disadvantage ship-owners.

We believe that IMO should be allowed to complete its work on CO_2 emissions from ships without any undue obstruction from regional institutions. What is important is the end result; the adoption of a meaningful international regulatory instrument that will provide for real emission reductions in a fair, non-discriminatory and orderly fashion.

The Prevention of Shipping Accidents

SOME COMMENTS ON THE ROLE OF INDUSTRY IN THE PREVENTION OF ENVIRONMENTAL HAZARDS RELATED TO SHIPPING

Aldo Chircop[1]

Shipping is by far the most energy efficient and cost-effective means of commercial transportation and lowest producer of marine and atmospheric pollution. Close to half of the 100,000 global ocean-going fleet moves cargo, but consumes only 2–4% of global annual fossil fuel.[2] At 2.7% of global emissions, shipping is a significant lower emitter of CO_2 than transportation by road, rail and air.[3] Even so, while the Joint Group of Experts on the Scientific Aspects of Marine Environmental Protection (GESAMP), an international advisory body sponsored by nine UN organizations, highlights that activities on land are the major source of marine pollution, including by far most emissions of greenhouse and ozone depleting gases affecting the oceans,[4] oil continues to enter the marine environment despite international regulation.[5]

As a major industrial activity, shipping produces significant threats to human health and the environment, and this impact ought to be considered cumulatively with other human induced impacts on the marine environment and human health.[6] Much of current international regulation for

[1] Director, Marine & Environmental Law Institute, Schulich School of Law, Dalhousie University, Halifax, Nova Scotia, Canada.

[2] James J. Corbett and James Winebrake, 'The Impacts of Globalisation on International Maritime Transport Activities: Past Trends and Future Perspectives', paper presented at the OECD/ITF Global Forum on Transport and Environment in a Globalising World 10–12 November 2008, Guadalajara, Mexico, 14 and 18 <www.oecd.org/dataoecd/10/61/41380820.pdf> accessed 13 January 2012.

[3] International Maritime Organization, 'Control of Greenhouse Gas Emissions from Ships Engaged in International Trade', Submission by the IMO to the United Nations Climate Change Conference, 8th Session of the Ad Hoc Working Group on Long-Term Cooperative Action (AWG-LCA 8), Fifteenth Conference of the Parties (COP 15), Copenhagen, 7–18 December 2009, 3.

[4] GESAMP, *A Sea of Troubles*, GESAMP Reports & Studies No. 70 (UNEP: Grid-Arendal, 2001) 19.

[5] GESAMP, *Estimates of Oil Entering the Marine Environment from Sea-based Activities* (London: IMO, 2007) 3 and 61 et seq.

[6] GESAMP (n. 3) 7 and 24. Oil pollution from ships and oil and gas activities 'has relatively minor effects on coastal and seabed ecosystems'. The report continues to state that 'Large-scale oil spills have had serious—if local and temporary—effects, but most of the

shipping addresses episodic pollution events, such as oil spills, dumping, sewage and oily water, whereas routine pollution such as atmospheric, exotic species, hull coatings and underwater noise are less addressed. Modern ship engines '... oxidize nitrogen effectively (thereby increasing NOx emissions), and emit many of the impurities of residual fuel (including sulphur, toxics, and heavy metals) out the ship stack.'[7] A study undertaken for the purposes of the then proposed North Environmental Emission Control Area estimated that without controlled emissions, premature mortality rates would be at between 5,100–12,000, in comparison to 3,700–8,300 with emissions controls.[8] Similarly: chronic bronchitis would be reduced from 3,500 to 4,600; acute bronchitis from 13,000 to 9,300; acute respiratory symptoms from 6.5 million to 3.4 million; hospital admissions from 8,400 to 3,500; and emergency room visits from 4,100 to 2,300.[9]

There is no doubt that the international regulation of shipping has had considerable success in reducing pollution of the marine environment, and indeed the international approach taken to regulate this industry has had far more success than other ocean industries, such as fishing and offshore oil and gas development and production. However, one cannot simply settle for 'shipping has done its fair share for protecting the environment, and other polluters need to pick up the slack.' On the contrary, shipping can continue to play a leadership role by providing a model of international regulation and by continuing to raise standards and address further those areas: (1) where we have gaps in existing international regimes, or (2) where we do not have gaps, but there is insufficient subscription by States, or (3) where there is no gap or lack of State subscription, but an existing or potential tool for environment protection is underutilized:

- Gaps: examples include greenhouse gas and particulate matter emissions from ships, currently under consideration at the International Maritime Organization (IMO), and underwater noise from ships as a marine pollutant, which is suspected to have adverse impacts on certain marine animals.

---

oil reaching the oceans comes from much less dramatic sources, such as routine discharges from ships, air pollution, and engine oil put down the drains'. Ibid.

7 Corbett and Winebrake (n. 1) 14.

8 Proposal to Designate an Emission Control Area for Nitrogen Oxides, Sulphur Oxides and Particulate Matter, Submitted by the United States and Canada, IMO Marine Environment Protection Committee, IMO Doc. MEPC 59/6/5, 2 April 2009.

9 Ibid.

- Insufficient subscription: with reference to existing conventions, examples here include ballast waters, antifouling, ship recycling and wreck removal conventions.[10]
- Underutilised tools: an example here is the Convention on Prevention of Pollution from Ships, 1973–78 (MARPOL) where there are marine regions around the world with still insufficient port reception facilities[11] or where no special areas are designated, despite their particular sensitivities to the pollutants addressed in the convention's annexes (e.g., South China Sea[12] and Arctic Ocean).[13]

The gist of my presentation may be protested by those that might say what I am pointing out are really responsibilities of States, not industry. While I agree that they are State responsibilities in a formal sense, that is, for the purposes of international law-making, I would counter argue that industry is affected, for better or for worse, by rules and standards. I hold that industry has the punch necessary to lobby governments into action and for providing the tools to do so.[14] They can do so directly or through the IMO, where numerous industry actors have very well-organized presence.

There are numerous examples how industry has played and continues to play a vital role in helping to establish and maintain safety and environmental standards for shipping that have most positive impact on maritime trade, such as in relation to the transportation of dangerous goods. Governments listen to key industry players, such as those that play instrumental roles in maritime trade. After all, behind every one percent of GDP growth in OECD countries in 1992–2006 there is four percent growth in trade, the majority of which is maritime.[15] Industry has always been at

10 For information on ratifications of IMO instruments see IMO, Status of Conventions <www.imo.org/About/Conventions/StatusOfConventions/Pages/Default.aspx> accessed 13 January 2012.

11 Noted by the IMO itself on numerous occasions. See for instance IMO, Port reception Facilities <www.imo.org/OurWork/Environment/PollutionPrevention/PortReceptionFacilities/Pages/Default.aspx> accessed 13 January 2012.

12 See Aldo Chircop, 'Regional Cooperation in Marine Environmental Protection in the South China Sea: A Reflection on New Directions for Marine Conservation' (2010) 41 Ocean Development & International Law 1.

13 As pointed out in Arctic Council, Arctic Marine Shipping Assessment 2009 Report <www.pame.is/amsa/amsa-2009-report> accessed 13 January 2012.

14 See for example the recent work of the International Shipping Federation and International Chamber of Shipping in promoting treaty ratification <www.marisec.org/Treaty%20Ratification%20Brochure.pdf> accessed 13 January 2012.

15 Corbett and Winebrake (n. 1) 7.

the forefront of technological development, and I would argue that they will retain that role. It is a leadership role that is crucial for providing the international community with key tools for marine environment protection and for persuading governments to scale up standards.

In conclusion, shipping has long experienced an international image of a major polluter. That image was deserved in the 1960–70s and continued to be earned in the 1980s and into the 1990s mostly as a result of major casualties at the centre of which was either a substandard ship or a badly-managed/owned ship, and occasionally as an unfortunate accident. But over time and especially since the turn of the millennium, good owners have incrementally cleaned-up ship management practices, shunning away bad ship owners, producing safer and cleaner transportation. Bad ship owning and management should not continue to colour an entire industry. The work of the Hellenic Marine Environment Protection Association, among sister organizations, deserves to be mentioned in showing the better side of industry.[16]

However, in a contemporary context the lingering polluter image persists in some respects, frequently forcing shipping into defensive postures. The message is that it takes time and effort to dispel an acquired image, but I would argue that through the exercise of a more proactive leadership role in enhancing environmental standards shipping would further contribute to the prevention of environmental hazards in a more visible manner, and thereby be well on its way to building a new image to replace the old.

[16] For an overview of their work, see HELMEPA <www.helmepa.gr/index2.php> accessed 13 January 2012.

DISCUSSION SUMMARY
WORKING GROUP: THE PREVENTION OF SHIPPING ACCIDENTS

Henrik Ringbom[1]

The discussion was divided in two main parts: firstly focussing on the adequacy and completeness of global regulation in shipping (with respect to law-making and enforcement); secondly, the nature and implications of EU's increasing activity in maritime safety legislation were discussed.

Shipping was recognised as being an effective and clean mode of transport with a good safety record. Lawrence Barchue noted the positive trend in the decreasing number of maritime accidents/casualties in the past 20 years. As far as *rule-making* is concerned, the panel agreed the scope of the maritime safety and environmental protection rules adopted by the IMO for international shipping is quite comprehensive and significantly more developed than in many other areas, such as fishing. It was considered that most relevant topics for this purpose are already regulated in the existing IMO conventions and that, in that sense, there are few outstanding 'gaps' in regulation. Barchue also stressed that in order to be useful, the existing regulations also need to be dynamic and adaptable to a changing world.

Aldo Chircop, however, pointed out that there are certain gaps in the area of air emissions which call for, if not new rules, at least a broader application of the existing ones. Apart from that, the most obvious inadequacy is that some of the conventions are not yet widely ratified and in force, such as the 2004 Ballast Water Management Convention or the 2009 Hong Kong Convention on Ship Recycling. He also considered that certain regions, such as the polar regions, sometimes may require special protection under the IMO Conventions, such as: emission standards under the MARPOL Convention; additional training requirements under the STCW Convention; or specific rules on traffic monitoring and collision regulation. Alexander Proelss agreed that matters relating to training of seafarers, and more generally to the 'human element', required continued attention and review, though not necessarily an entirely new set of rules.

[1] Dr. Henrik Ringbom, Head of Unit, European Maritime Safety Agency.

The Chair raised the question as to whether IMO is generally speaking more successful in adopting technical rules for ships than operational issues, which relate to the activities by the ship's crew on board the ship. Reference was made to the recent experience with regulating greenhouse gases. Out of the package of measures IMO had decided to explore in this area, the regulatory attention had almost exclusively been on the energy efficiency design index, which was a measure to be implemented on the design phase of (future) ships. By contrast, operational measures, which could apply to any ship, have been deemed to be unsuitable for mandatory requirements at the present time.

Audience member Kai Trümpler, Federal Maritime and Hydrographic Agency (BSH), asked whether the safety issues related to ships and offshore wind farms would deserve some additional attention. Andreas Chrysostomou, also in the audience, responded by referring to various ongoing activities on this matter within the IMO's Maritime Safety Committee. He also noted that IMO cannot address all aspects of safety regulation and that some issues, including operational matters, could well be handled by States, individually, bilaterally or within a region. With respect to seafarers, Chrysostomou considered that the social aspects of the profession deserve more regulatory attention, including the interaction between man and machine on board ships and issues related to monotony and boredom of crew members.

It was concluded that the main challenges today, in terms of regulatory 'gaps' in shipping seem to lie in the 'softer' part of regulation, i.e. rules that focus on the human element, training of crew, social aspects, environmental awareness etc.

Turning to *compliance and enforcement*, it was initially noted that the existing regulatory system for shipping is largely based on a flag State approach, in which the prime responsibility for implementing and enforcing the international safety and environmental rules lies with the State whose flag the ship is entitled to fly. This responsibility of flag States emanates not only from provisions in the individual IMO conventions but also from UNCLOS, which includes a number of articles on the minimum obligations of flag States in the field of safety and environmental protection. Port States' increasing role to verify compliance during the past decades was considered to be a very welcome development. However, while port State control may complement the role of flag States in verifying compliance, and put additional pressure on flag States to live up to their obligations, port State control does not, and should not, replace the primary responsibility of flag States.

Barchue was asked to comment upon the effect of the IMO's on-going voluntary audit scheme for Member States on flag State implementation, considering the possibility that this scheme may be transformed into a mandatory instrument for flag States in the future. He responded that to some extent accidents are unavoidable. Ships are mechanical structures and with all such structures there will be failures from time to time. Control, procedures and inspections, no matter how sophisticated, will be unable to detect all potential risks with a ship. The ambition of zero-accidents was therefore not a realistic one. A very welcome move to incite and improve flag State implementation was the step taken by some port State control authorities, notably the Paris MOU, to record and rate flag State performance and link that performance to the likelihood of inspection by individual ships based on their flag. Barchue also noted that as long as UNCLOS remains as it is, non-flag States will have to rely on quite cautious mechanisms to regulate (foreign) ships, in contrast to aviation, for example, where enforcement can be linked to refusal of overflight or landing permits.

Proelss agreed, but noted that the obligations imposed on flag States in UNCLOS are both numerous and detailed in nature. So there is no real question at this point as to *whether* flag States are subject to the international safety and environmental standards, but rather who is best placed to ensure that these rules are complied with. He thought port States, and in particular regional bodies such as the EU, may be of key importance in this regard.

With that observation the discussion shifted to the *role of the EU* in maritime safety and environmental regulation. It was noted that the absence of focus by IMO on the role of coastal and port States has provided the opportunity for individual States and regions to emphasize that role in enhancing the regulatory system. This was very true of the EU which almost exclusively had focused on measures to be taken by the Member States in their capacity as port and coastal States. Due to concerns related to a 'level playing field' in a global industry, measures targeting the EU Member States only in their capacity as flag States had been quite uncommon until now.

The Chairman suggested that the development of the EU measures over time is one of growing independence from the international (IMO) rules. Mainly through measures implemented for any ship calling at an EU port, but in some cases also for ships only passing through the coastal waters of Member States, a gradual development has taken place in which the EU, while still mostly implementing the international rules, tended to add more and more 'unilateral' elements to its rules over time. This increasing independence could raise questions of international law. He then asked the panellists whether they thought this trend was a positive development in terms of more effective regulation of safety and the environment or if it was one which should be cautioned against for legal, policy or other reasons.

Generally speaking, Chircop considered that the development within the EU was a positive one. He further noted that under international law entry into ports is a privilege rather than a right. It can accordingly be denied. Yet, there are cases where it may be transformed into a right, for example through trade agreements or other instruments that provide for access. Another exception is where the ship is in distress, which may give rise to the rule of customary law that ships have a right to enter a place of refuge or a safe haven.

It was noted that one of the differences between concerted EU 'port State' measures, as opposed to requirements by individual Member States had to do with size. Individual States would usually refrain from enacting strict entry into port requirements on ships for economic reasons. Ship operators may simply decide to divert their ships to ports of a neighbouring State instead. In a block of 27 Member States this risk of diversion is significantly reduced, which in part may explain the difference in substance between EU's and individual Member States' policies in shipping.

Proelss agreed with this assessment and added that the EU could also serve a useful role in clarifying certain rules of the law of the sea that are not clearly laid down in the regulatory framework. He cautioned, however, against overstating its importance in this regard. Many of the relevant rules represent international customary law and cannot be altered by the EU alone. Great care is needed by the EU to ensure that legal limits are not overstepped. In terms of result, he agreed that the effects of the EU's activities in maritime safety and pollution prevention had been positive until now, but that in itself should not justify ignorance of international law. The tendency appears to have been to test the limits of international law and in the longer run such an approach may erode the respect for, and validity of, international law.

Chrysostomou in the audience agreed that EU maritime legislation may at times affect and even serve to develop international law in this field. In fact testing the limits of international law may be understandable as it may sometimes be the only way to advance it. But it should not be forgotten that the overwhelming part of EU legislation is aimed at global harmonisation and often actually implements IMO rules. Sometimes international provisions are even copied directly into EU directives and regulations, hence bringing them into the sphere of EU law and making them binding for Member States even without ratification.

Trümpler in the audience also agreed on the legal possibility to place access conditions for foreign ships' entry into port, but wondered about whether the right issues were being addressed by the EU. Since most accidents seem to be attributable to the 'human element', and since early human intervention could nearly always have prevented an accident from happening in the first place, perhaps more attention should be given to issues related to seafarers and labour law matters.

The Chair responded by giving some examples where the EU and EMSA have been active in regulating and enforcing rules on seafarers' training, social matters and investigation of accidents.

In the last round of comments Barchue highlighted the need to take into account that different States have very different needs and interests in this field. In order to be effective, the maritime safety rules need to accommodate all those diverging interests.

Chircop came back to the issue of seafarers and noted that the changes that the industry had undergone in the past decades had had significant consequences for the seafaring profession. He questioned whether those changes had been reflected in the existing regulatory framework for seafarers, notably the training requirements.

Proelss returned to the balance between being ambitious and going too far, in relation to the EU's regulatory role in this field. He agreed with Chrysostomou that the majority of EU rules support and give effect to IMO rules, but estimated that a small part (the figure of 2 per cent was singled out as a rough estimate) of the EU rules nevertheless could be problematic from the point of view of international law.

The Chair concluded that the importance of both global and regional rules seemed to be recognised in the discussions. The quantity or breadth of regulation was not perceived to be a shortcoming, whether at IMO or EU-level, but for both organisations some doubts had been raised as to whether the right issues were currently regulated and whether they were regulated in an optimal way. A particular difficulty that had been highlighted throughout the discussion was to address the 'softer' parts of maritime safety, notably everything related to crews and training, and more generally operational requirements.

With respect to the increasing regional activity in regulation of shipping, which has traditionally been subject to uniform, global rules, the Chair noted that the overall sentiment in the panel seemed to be that the EU's activity in this field was supported and considered to add value to maritime safety and the protection of the environment. He wondered whether this response was perhaps due to the absence, so far, of EU rules that clearly conflict with international law. While certain tensions between EU and global rules in this area had been noted during the discussion, no mention had been made of cases where the line had been crossed in which an EU act aimed at implementing, exploring, clarifying or even developing international law had been found to breach it. In that sense it would seem that the EU had indeed, as recommended in the discussions, made an effort to ensure that its regional rules on shipping are compatible with IMO conventions and the law of the sea.

PART FOUR

PROTECTION OF THE MARINE ENVIRONMENT AND CLIMATE CHANGE

THE IMPACT OF GLOBAL WARMING ON THE OCEANS

Mojib Latif[1]

1. Abstract

Twentieth century climate exhibits a strong warming trend. There is a broad scientific consensus that the warming contains a significant contribution from enhanced atmospheric greenhouse gas (GHG) concentrations due to anthropogenic emissions. The climate will continue to warm during the 21st century due to the large inertia of the Earth System and in response to additional GHG emissions, but by how much remains highly uncertain. This is mainly due to three factors: natural variability, response uncertainty, and GHG emission scenario uncertainty. Uncertainty due to natural variability dominates at short time scales of years up to a few decades, while at the longer centennial time scales scenario uncertainty provides the largest contribution to the total uncertainty. Yet the likelihood is large that globally averaged surface air temperature (SAT) will exceed a level toward the end of the 21st century that will be unprecedented during the history of mankind, even if strong measures are taken to reduce global GHG emissions. It is this long-term perspective that demands immediate political action.

2. Introduction

The atmospheric carbon dioxide (CO_2) concentration has strongly increased since the start of industrialization (Figure 1) in response to anthropogenic emissions and reached a level which is unprecedented in man's history. The present CO_2 concentration amounts to about 390ppm as opposed to the pre-industrial concentration of 280 ppm. Carbon dioxide is a greenhouse gas and known to warm the Earth's surface, as it is transparent for the short-wave solar but not for some of the long-wave infrared radiation emitted by Earth's surface. The globally averaged surface air temperature (SAT) of the planet has warmed by about 0.7°C during the 20th century

[1] Helmholtz Centre for Ocean Research and Kiel University, Germany.

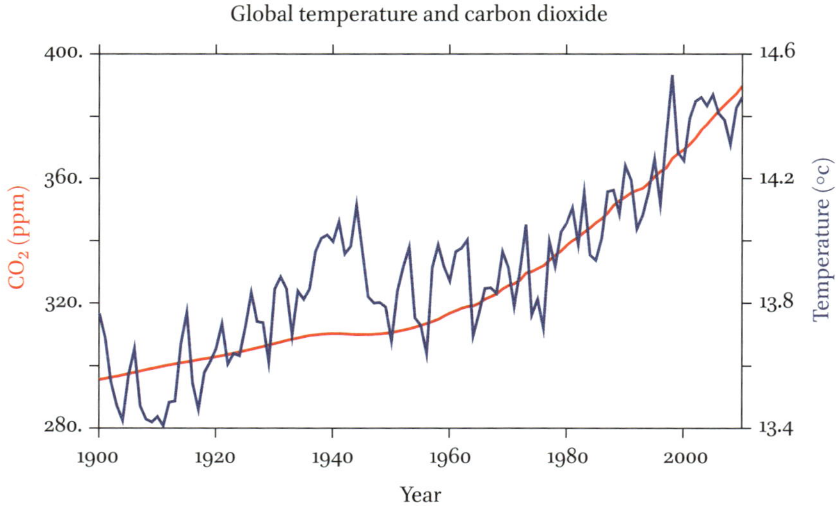

Figure 1. Observed annual globally averaged SAT (°C, blue line) and smoothed atmospheric carbon dioxide concentration (ppm, red line) during 1900–2008.

(Fig. 1), global sea level has risen by just under 20cm, and many mountain glaciers and Arctic sea ice have considerably retreated. There is compelling scientific evidence that at least half of 20th century warming was forced by the increase of GHG concentrations.[2] They will continue to rise over the next years and possibly even decades, which together with the inertia of the climate system will support further global warming during this century. A global mean temperature rise implies higher warming over land than over oceans, with the tropical regions warming least and the northern polar region warming the most.

3. Natural Variability

However, climate change predictions are inherently uncertain. This does not come as a surprise to climate scientists. The media and public are generally confused by uncertainty. Hawkins and Sutton[3] estimated the

---

[2] IPCC, 'Summary for Policymakers' in S. Solomon, D. Qin, M Manning, Z. Chen, M. Marquis, K.B. Averyt, M. Tignor and H.L. Miller (eds) *Climate Change 2007: The Physical Science Basis. Contribution of Working Group I to the Fourth Assessment Report of the Intergovernmental Panel on Climate Change* (Cambridge University Press 2007).

[3] Ed Hawkins and Rowan Sutton, 'The Potential to Narrow Uncertainty in Regional Climate Predictions' (2009) 90 Bulletin of the American Meteorological Society 1095–1107 <http://journals.ametsoc.org/doi/pdf/10.1175/2009BAMS2607.1> accessed 03 July 2012.

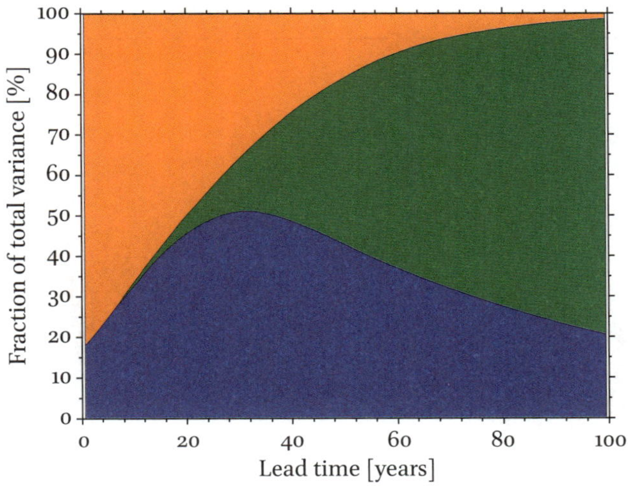

Figure 2. The fraction of total variance in annual mean surface air temperature predictions explained by the three components of total uncertainty (orange: internal variability, blue: model uncertainty, and green: scenario uncertainty).

contributions to the total prediction uncertainty from natural internal variability,[4] response (model) uncertainty, and scenario uncertainty. The relative contributions from internal variability and response uncertainty dominate at short lead times up to a few decades (Figure 2). At longer lead times of several decades, scenario uncertainty provides the largest share to prediction uncertainty. Furthermore, uncertainty generally increases with smaller space scale. It is important to note, however, that the contributions to prediction uncertainty from internal variability and especially from response (model) uncertainty are potentially reducible through progress in climate science.

Surface air temperature (SAT) during the 20th century displays a gradual warming and superimposed short-term fluctuations, while carbon dioxide evolved rather smoothly. The warming trend contains the climate response to enhanced atmospheric GHG levels but presumably also a natural component. The temperature ups and downs around the trend which are particularly pronounced in the Arctic (Figure 3) largely reflect natural variability. Natural climate variations are of two types, internal and external. Internal variability is produced by the climate system itself

[4] We can categorize the natural climate variations into two classes: internal and external. Internal fluctuations are produced by the climate system itself. External fluctuations on the other hand need a forcing, a change in the boundary conditions (see below).

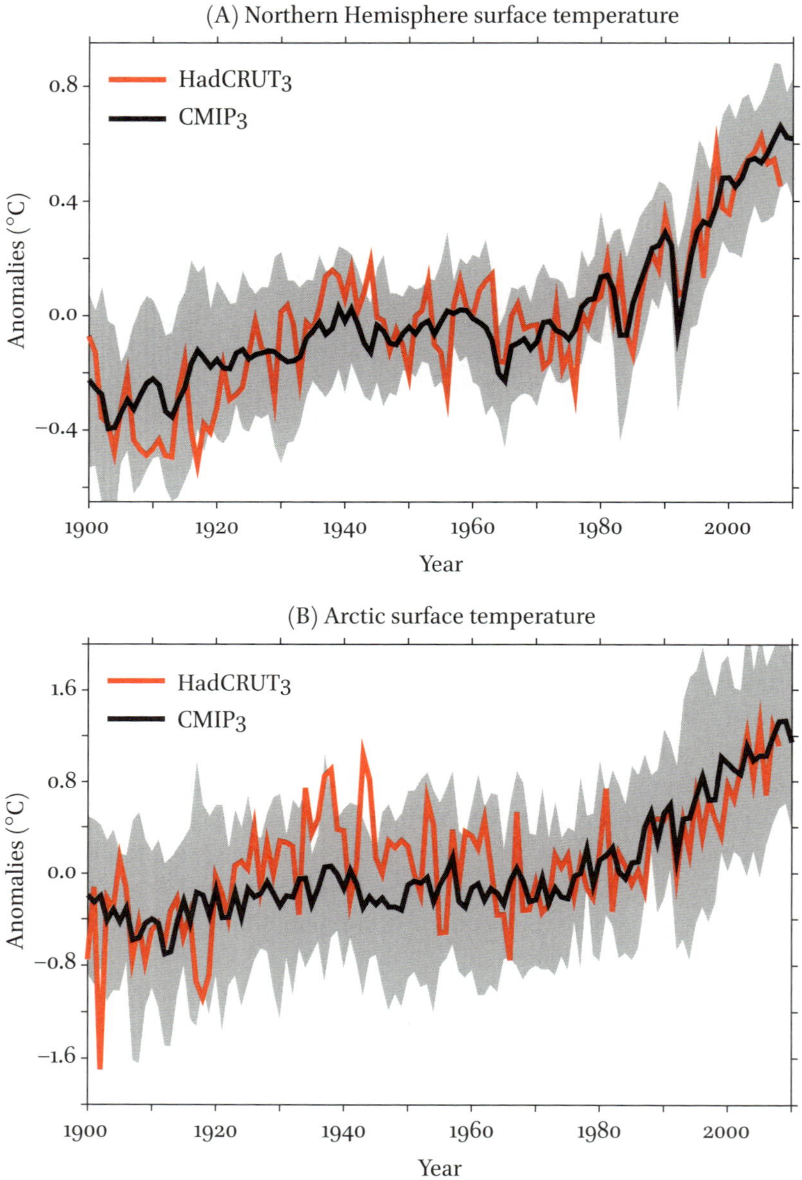

Figure 3. (A) Observed Northern Hemisphere averaged and (B) Arctic (60–90°N) SAT (red lines) and the multi-model mean of the CMIP3 models. The shading displays the model spread.

due to its chaotic nature. External fluctuations need a forcing, a change in the boundary conditions. Volcanic eruptions and fluctuations in solar output are examples. The eruption of the Philippine volcano Mt. Pinatubo in 1991, for instance, caused a relatively short-lived drop in global SAT of about 0.2°C in 1992 (Figure 1); and an increase of the solar radiation reaching the Earth may have contributed together with other processes to the mid-century warming (MCW) during 1930–1940 (Figure 1). The anthropogenic influence on climate is also considered as external.

One way to estimate the external contribution to the 20th century SAT change is to run climate models with all (known) observed external (natural and anthropogenic) forcing in ensemble mode with different initial conditions. Figure 3 shows such simulations of Northern Hemisphere (upper panel) and Arctic (lower panel) SAT. The average over all (IPCC) models taken from the CMIP3 database[5] is sometimes referred to as the 'consensus' (black lines) and a measure of the externally driven climate change. The ensemble average displays a clear gradual upward trend in both indices that is consistent with the observed trend, but fluctuations are strongly damped by the averaging. Different initial climate states yield different realizations of internal variability even under identical external forcing, one reason for the spread (as denoted by the gray shading), as integrations are performed in ensemble mode with different start conditions.

A well-known example of such internal variability on interannual time scales is El Niño, a warming of the Equatorial Pacific occurring on average about every 4 years which is the warm phase of the El Niño/Southern Oscillation (ENSO). The record event of 1997/1998 'helped' to make 1998 the warmest year to date globally (Figure 1).[6] The year 2009 also happened to be an El Niño year, which supported, for instance, a weak Atlantic hurricane activity, as El Niño causes enhanced upper-level vertical wind shear over the Tropical Atlantic which is known to hinder hurricane development. The event which was considerably weaker that the 1997/1998 event persisted into 2010 and was partly responsible for the period January–June 2010 being the warmest first half year on record globally. However, the subsequent major cooling in the Equatorial Pacific referred to as La Niña 'compensated' the El Niño warming to some extent so that 2010 did not become the new record year. This discussion illustrates

[5] Gerald A. Meehl et al., 'The WCRP CMIP3 Multi-Model Dataset: A New Era in Climate Change Research' (2007) 88 Bulletin of the American Meteorological Society 1383.

[6] According to the SAT analysis of the British Meteorological Office.

that caution should be exercised when interpreting short-term climate trends that are based on only several months or a few years.

Decadal-scale variability is also evident in the data.[7] In Figure 3, the deviation of the observed temperature evolution from the relatively smooth 'consensus' reflects the internal variability, assuming the model-mean is a reliable estimate of the externally-forced signal; and multidecadal changes are obvious in the residuals. Such multidecadal or even longer time scale natural variability, internally or externally driven, may mask anthropogenic climate signals which evolve on similar time scales. It has been concluded,[8] for instance, that the expected anthropogenic weakening of the Meridional Overturning Circulation (MOC),[9] a prominent circulation in the Atlantic Ocean transporting large amounts of heat northward thereby contributing to the mild climate of Northern Europe, may not be detectable during the next decades due to the presence of strong internal multidecadal variability. This may not only apply to the MOC itself but also to other potentially related quantities such as surface air temperature in parts of Europe and North America, Sahel rainfall or Atlantic hurricane activity which are also characterized by pronounced multidecadal variability and may hamper early detection of an anthropogenic signal.

To some extent, we need to 'ignore' the natural fluctuations, if we want to 'see' the human influence on climate. Had forecasters extrapolated the MCW into the future, they would have predicted far more warming than actually occurred. Likewise, the subsequent cooling trend, if used as the basis for a long-range forecast could have erroneously supported the idea of a rapidly approaching ice age. The scientific challenge is to quantify the anthropogenic *signal* in the presence of the background climate *noise*. The detection of the anthropogenic climate signal thus requires at least the analysis of long records, because we can be easily fooled by the short-term natural fluctuations, and we need to understand their dynamics to better estimate the *noise level*. Sophisticated fingerprint methods[10]

---

[7] E.g. Chunzai Wang and Shenfu Dong, 'Is the Basin-Wide Warming in the North Atlantic Ocean Related to Atmospheric Carbon Dioxide and Global Warming?' (2010) 37 Geophysical Research Letters.

[8] E.g. Mojib Latif et al., 'Is the Thermohaline Circulation Changing?' (2006) 19 Journal of Climate 4631.

[9] The MOC is often referred to as the Gulf Stream Circulation.

[10] G. Hegerl et al., 'Detecting Greenhouse Gas Induced Climate Change with an Optimal Fingerprint Method' (1996) 9 Journal of Climate 2281.

maximizing the *signal-to-noise ratio were* applied to detect the anthropogenic signal in observations.

The components of this strategy include observations, information about natural climate variability, and a model-derived 'guess pattern' representing the expected time-space pattern of anthropogenic climate change. The expected anthropogenic climate change is identified through projection of the observations (say during the 20th century) onto the (model-derived) fingerprint. The latter can be optimized by weighting those components more strongly which are less 'inflated' by natural variability. Furthermore, the relative contributions of different external drivers of climate have been quantified. The results appear to be sufficiently robust to conclude that the observed climate change during the last decades cannot be explained solely by natural variability and is consistent with a combined greenhouse gas and aerosol[11] forcing, but inconsistent with greenhouse gas or solar forcing alone.[12]

4. Response Uncertainty

The spread seen in Figure 3 also reflects response (model) uncertainty, as different models simulate different climate responses even when forced by the same GHG concentration or emission scenario.[13] Climate models are grounded on basic physical principles. As such they are fundamentally different to empirical models which are used, for instance, in economic forecasting. Climate models, however, are far away from being perfect. Errors in annual mean SAT, for instance, can amount to several degrees in some regions.

Limitations in computer resources dictate the use of either reduced or relatively coarse-resolution models. As a consequence many important processes cannot be explicitly simulated; they must be *parameterized*. Some processes like cloud formation, cloud-radiation interaction and its influence on the general circulation of the atmosphere, or the role

[11] These are small particles which are also produced by the combustion of fossil fuels and generally exert a net cooling effect on the Earth's surface.

[12] G. Hegerl et al., 'Multi-Fingerprint Detection and Attribution Analysis of Greenhouse Gas, Greenhouse Gas-Plus-Aerosol and Solar Forced Climate Change' (1997) 13 Climate Dynamics 613.

[13] Emissions are prescribed if the climate model carries a carbon cycle module which calculates the fraction that remains in the atmosphere.

of mesoscale eddies[14] for the large-scale ocean circulation are not completely understood and major sources of model bias. The large model biases should be kept in mind when it comes to regionalization of climate change projections. Global climate models, however, can be potentially improved and thus uncertainty stemming from model error reduced.

The *climate sensitivity* is defined as the equilibrium change in globally averaged SAT in response to a doubling of the pre-industrial atmospheric CO_2 concentration (i.e. from 280 to 560ppm). IPCC AR4 stated that the value '... is likely to be in the range 2°C to 4.5°C with a best estimate of about 3°C, and is very unlikely to be less than 1.5°C'. In the IPCC definition *likely* refers to an outcome or result when its likelihood is greater than 66% probable. *Very unlikely* means a probability of less than ten percent. Thus there is a non negligible probability that the climate sensitivity is either considerably smaller or larger than the best estimate of 3°C. Obviously, just communicating the best estimate to the public (obtained by averaging over the ensemble of climate models) is inappropriate.

The uncertainty in climate sensitivity itself is a good reason to demand reductions of GHG emissions, because the possibility of 'a dangerous anthropogenic interference with the climate system'[15] cannot be ruled out with high confidence. That is we do not really know at which levels atmospheric GHG concentrations should be stabilized. Society may be lulled into a false sense of security by presenting smooth model projections of climate change. In some regions anthropogenic forcing on the climate system could kick start abrupt and potentially irreversible changes. For these sub-systems the term 'tipping element' is commonly used.[16] Warming over the Greenland ice sheet, for instance, accelerates ice loss from outlet glaciers and lowers ice altitude at the periphery, which further increases surface temperature and ablation. A complete meltdown would result in a rise of sea level of up to 7m globally. The exact tipping point for disintegration of the ice sheet, however, is unknown, since current models cannot capture the complex processes accurately.

---

14 Eddies are the analogues of the mid-latitudinal storms in the atmosphere, but they are much smaller in size.

15 United Nations Climate Convention, Rio de Janeiro (1992).

16 Timothy M. Lenton et al., Tipping elements in the Earth's climate system (2008) 105 Proceedings of the National Academy of Sciences of the United States of America 1786.

Similarly, global warming can potentially affect the stability of the West Antarctic Ice Sheet,[17] with ice volume equivalent to 5m of sea level. Protective ice shelves could shrink or even disappear and the ice sheet would become vulnerable to melting. Again, the precise tipping point for complete disappearance of the ice sheet is highly uncertain.

5. Climate Change Projections

To predict the future climate we have to consider both natural variability and anthropogenic forcing. The latter is taken into account by assuming scenarios about future GHG and aerosol emissions. The scenarios (Figure 4) cover a wide range of the main driving forces of future emissions, from demographic to technological and economic developments. IPCC AR4 published only climate projections based on several scenarios with no attempt to take account of the likely evolution of the natural variability. This by definition yields relatively smooth trajectories if the results are averaged over many realizations and models, as discussed above. In the real world, the natural variations will introduce a large degree of irregularity (Figures 1 and 3), and even short-term cooling may occur over the next years. This could have been explained better to the public, as in some media reports the existence of global warming has been questioned after for more than ten years no global SAT record has been observed. Had we emphasized more the uncertainty, that debate which confused many people could have been avoided. Albert Einstein once said that we should make 'things as simple as possible, but not simpler'.

As can be clearly inferred from Figure 4, it is the long-term strategy that we adopt which matters on the long time scales of many decades. Because of the long residence time of CO_2 of more than 100 years in the atmosphere, climate response is governed by cumulative rather than current CO_2 emissions. Important is not the detailed emissions path, but that emissions are strongly reduced over a period of 50 to 100 years.[18] In order to stabilise the concentration of greenhouse gases in the atmosphere, GHG emissions would need to peak and decline thereafter. The lower

[17] David Pollard and Robert M. De Conto, 'Modelling West Antarctic Ice Sheet Growth and Collapse Through the Past Five Million Years' (2009) 458 Nature 329.

[18] Klaus Hasselmann et al., 'The Challenge of Long-term Climate Change' (2003) 302 Science 1923.

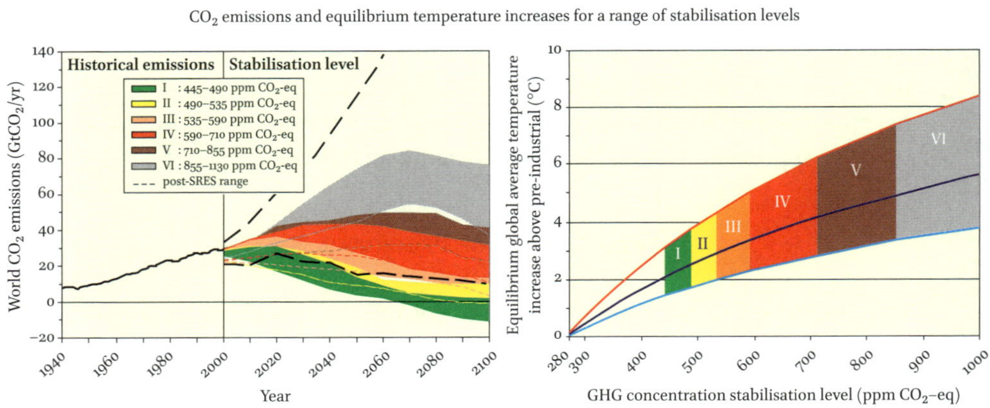

Figure 4. Global equivalent CO_2 emissions (Gt CO_2/yr) for 1940 to 2000 and emissions ranges for categories of stabilization scenarios from 2000 to 2100 (left panel); and the corresponding relationship between the stabilization target (ppm) and the likely equilibrium global average temperature increase (°C) above pre-industrial levels (right panel). Colored shadings show stabilization scenarios grouped according to different targets (stabilization category I to VI). The right-hand panel shows ranges of global average temperature change above pre-industrial levels, using (i) 'best estimate' climate sensitivity of 3°C (black line in middle of shaded area), (ii) upper bound of likely range of climate sensitivity of 4.5°C (red line at top of shaded area) (iii) lower bound of likely range of climate sensitivity of 2°C (blue line at bottom of shaded area). Black dashed lines in the left panel give the range of recent baseline scenarios published since the SRES.[19, 20]

the stabilisation level, the more quickly this peak and decline would need to occur. Stabilisation at lower concentration and related equilibrium temperature levels advances the date when emissions need to peak and requires greater emissions reductions by 2050.

Climate sensitivity obviously is a key uncertainty for mitigation scenarios that aim to meet specific temperature levels, for instance the 2°C-target that is limiting global warming to 2°C by the end of the century above pre-industrial levels. The timing and level of mitigation to reach a given temperature stabilisation level is earlier and more stringent if

[19] IPCC, 'Summary for Policymakers. Special Report Emission Scenarios' (2000).
[20] IPCC, 'Climate Change 2007: Synthesis Report. Contribution of Working Groups I, II and III to the Fourth Assessment Report of the Intergovernmental Panel on Climate Change' (2008).

climate sensitivity is high than if it is low.[21] Moreover, if the CO_2 concentration continues to strongly rise during this century, it will stay rather high throughout the millennium, as the removal of CO_2 from the atmosphere is a very slow process which is governed by the slow exchange between the surface and the deep ocean.

It is noted that approaching equilibrium can take several centuries, especially for sea level and scenarios with higher levels of stabilisation. Thermal expansion would continue for many centuries after GHG concentrations have stabilised, for any of the stabilisation levels assessed, causing an eventual sea level rise much larger than projected for the 21st century. If GHG concentrations had been stabilised at year 2000 levels, thermal expansion alone would be expected to lead to further sea level rise on the order of about half a meter over the next centuries. The eventual contributions from the Greenland and West Antarctic ice sheets is highly uncertain but could amount to several metres by the end of the millennium, and larger than from thermal expansion, should warming in excess of about 2°C above pre-industrial levels be sustained over many centuries.

6. Conclusions

What are the implications of all this? First, uncertainty is an integral part of climate change projections. It arises from natural variability, model and scenario uncertainty, and from our incomplete knowledge of the Earth System dynamics. Impact studies and political decisions to mitigate climate change have therefore to be based on inherently uncertain climate change projections for the 21st century. It is important to communicate this to the wider scientific community and the public.

However, although the exact global warming to be expected by the end of this century cannot be predicted with great accuracy, there is a very high likelihood that globally averaged surface air temperature (SAT) will rise to a level which will be unprecedented during the history of mankind. Global SAT during the last major warm period, the Eemian Warm Period, peaked about 125,000 years ago and was about 0.5–1.0°C higher relative to present levels. We can expect an additional warming during the 21st century of 0.3–0.9°C assuming year 2000 constant GHG concentrations just

21 Ibid.

due to the inertia of the Earth System, with a best estimate of 0.6°C (IPCC, 2007).[22] Furthermore, we cannot expect global GHG emissions to rapidly decline soon given their dramatic increase during the last two decades, which will support further global warming.

Improved global models will enable more reliable regional climate projections, in which society is mostly interested in. The Earth System exhibits a wide range of dynamical phenomena with associated physical, biological and chemical feedbacks that collectively result in a continuum of temporal and spatial variability. A new paradigm is emerging that of the *seamless prediction*.[23] In a nonlinear system such as the Earth System many time and space scales interact with each other. Certain weather patterns, for instance, such as blocking events which often trigger heat waves are not well captured in state-of-the-art climate models. It is thus questionable whether current global climate models can reliably predict regional precipitation changes in response to increased GHG concentrations, if they do not realistically simulate weather phenomena like blocking. Utilization of the insights gained from operational weather and seasonal prediction, and of the synergy between the weather and climate prediction communities is essential for the development of next-generation prediction systems.

Finally, we need to advance our understanding of how the Earth's biogeochemical cycles, including human actions, interact with the climate system by constructing Earth System Models (ESMs). 'Pure' climate models are based on an atmospheric circulation model coupled with an oceanic circulation model, with representations of land and sea ice dynamics. These models have been devised by climate modelers to study weather, climate, and potential changes, both natural and anthropogenic. An ESM adds, to all of these components, the interactive carbon cycle, and associated chemical and ecological tracers, which determine nutrients, plant biomass and productivity. The ESM also captures numerous types of emissions, variations of land surface albedo due to both natural vegetation changes and land use history, including agriculture and forestry, and aerosol chemistry. Adding these different components to the ESM represents a major step towards simulating the Earth's ecological systems. The vision of global modeling is an integrated ESM, projecting not only climate

22 IPCC (n. 2).

23 T.N. Palmer, F.J. Doblas-Reyes, A. Weisheimer, M.J. Rodwell, 'Seamless Prediction: Calibration of Climate Change Projections Using Seasonal Forecasts' (2008) 89 Bulletin of the American Meteorological Society 459.

variability on seasonal to centennial timescales, but also biogeochemical and ecosystem cycling and biological feedbacks on the climate system. It will allow, for instance, the study of coastal ocean ecosystems and is an important step toward ecological prediction. This is a comprehensive effort, requiring incorporation of climate dynamics, biogeochemistry, ecological processes and human activity.

USE AND PROTECTION OF THE SEAS IN TIMES OF CLIMATE CHANGE

Jochen Flasbarth[1]

Ladies and Gentlemen,

I am glad to have the opportunity to deliver a speech today at your environmental law conference, addressing you on behalf of the German Federal Environment Agency which has been actively committed to protecting the marine environment for decades.

1. Introduction

The hunger for raw materials by a world population which is set to grow to 9 billion by 2020, and increased technical possibilities have been key factors for the growing use and exploitation of our seas and its consequences for the marine environment. Increasingly, user interests are clashing with one another—I am thinking here mainly of fisheries, navigation, energy production, but also of agriculture—and with the 'interests' of marine protection, especially the protection and preservation of biodiversity. In addition to direct uses of the seas, land-based consumption of fossil fuels is leading to climate change, resulting in an increase in water temperatures and acidification phenomena in the world's oceans. Ecological impacts on important members of the marine food web are already visible today. That is why environmentalists have been calling for further regulatory intervention by governments for many years, all the more so as climate change is increasingly impacting on all the world's oceans.

In my speech, I shall deal with the tension between protection and use of the seas, whose important functions include the mitigation of man-made climate change and its consequences. Let me give you a brief outline of this area of tension.

First of all, it should be noted that healthy marine ecosystems make a significant contribution to mitigating climate change and its impacts,

[1] President of the German Federal Environment Agency.

thanks to their function as a carbon sink and their resilience. On the other hand, certain uses of the seas, such as offshore wind power, the envisaged sub-seabed storage of CO_2 and ocean fertilization, currently under debate, which are intended to help mitigate climate change, have the potential to adversely affect marine ecosystems in different ways. The major challenge we now face is to provide these uses with adequate 'ecological guard-rails' in order to ensure that they do not run counter to the protection of our most important carbon sink—the seas—while at the same time curbing greenhouse gas emissions.

I will first address the role of healthy marine ecosystems in climate change mitigation and then discuss the different uses of the seas and their potential impacts before moving on to the legal aspects of the protection and use of the seas.

2. How Can Protection and Uses of the Seas Stabilise the Climate?

2.1. *Preserving the Function of Our Seas as a Carbon Sink through Marine Protection and Climate Change Mitigation*

If we ask what role the seas play in regard to climate change we first need to realize that the oceans are our planet's biggest and most sustainable carbon sink. They can store about 50 times as much carbon as the atmosphere, about 2 gigatonnes (Gt) per year, and currently have already absorbed 30 percent of the greenhouse gas CO_2. This is referred to as 'blue carbon' to distinguish it from the 'green carbon' which is stored in land vegetation.

Most of the carbon is bound in the deep sea. It reaches these depths through the 'biological pump', the downward transport of dead organic material. CO_2 from the atmosphere dissolves in surface water and is taken up by minute algae, the phytoplankton, during photosynthesis and bound in biomass in the light-exposed water strata. Some of the biomass is metabolised to CO_2 and water by the phytoplankton and the remainder is available to consumers, for instance zooplankton, as food. When phytoplankton and zooplankton die, they sink to greater depths. The biological pump currently transports 10 to 11 Gt of carbon per year to the oceans' depths. Some of the carbon is again made available to other organisms during its downward transport whilst another part reaches even greater depths and will then be removed from the carbon cycle for a period of 500 to 1000 years, bound, or sequestered, at those depths. The minimum depth at which such sequestration takes place depends on the hydrographical conditions and varies between 200 to 1,000 metres (m). Only a very small proportion of carbon, about 0.2 Gt per year, is permanently sequestered in ocean sediments.

Not only the deep sea and ocean sediments can store carbon, but also coastal ecosystems have considerable capacity for carbon storage. In particular, salt meadows, seagrass meadows and mangrove forests, which cover 0.5 percent of total global marine area, currently store about 0.1 Gt of carbon per year, which is more than half of the carbon sequestered in ocean sediments. However, the storage function of these important carbon sinks is under threat. Globally, human activity has already destroyed one third of seagrass meadows, a quarter of salt meadows and 35 percent of mangroves.

We can therefore state that the preservation of healthy marine ecosystems is essential to the mitigation of man-made climate change. Efforts should focus on both protecting natural coastal ecosystems and preserving intact marine food webs, which are vital for the functioning of the biological pump. Protection of the seas therefore also means protection of the climate.

2.2. *Strengthening the Resilience of Marine Ecosystems to Mitigate the Impacts of Climate Change on the Seas*

Strengthening the resilience of marine ecosystems is important in two respects: Firstly, for reducing the negative impacts of climate change on the seas, and secondly, for ensuring that they can continue to be used in a sustainable manner.

In the future, climate change will place considerable additional pressures on marine ecosystems, especially where prior damage exists. To give an example, a coral reef which is already damaged—as a result of illegal fishing with poison or dynamite or fishing with (legal) bottom trawls—will be particularly vulnerable to exceptionally strong heat waves. There is evidence that warming promotes what is known as coral bleaching, which may cause the death of coral reefs. Coral bleaching shows that even the healthiest reefs are not immune to the impacts of climate change. It was found, however, that reefs affected by coral bleaching have a greater chance of survival when they are healthy and located in pristine regions. Therefore, in addition to taking measures to mitigate climate change, it makes sense to strengthen the resilience of coral communities through protection.

A second example: A fish species whose stocks have been severely depleted by overfishing will recover at a much slower pace if coastal ecosystems in the form of mangroves, salt meadows or seagrass serving as 'nurseries' for this species are affected by infrastructural measures or pollution. Warming is an additional stress factor. Therefore, in the next years it will be crucial to curb current overfishing so that climate change will meet with sufficiently resilient ecosystems.

2.3. *The Seas as an Inexhaustible Energy Source—Wind Power, Wave Power*

Having described how protection of the seas can help to combat climate change, let me now turn to the uses of the seas and their contribution to climate protection.

Offshore wind energy will make a significant contribution to energy supply in the future. As winds are stronger and steadier at sea than on land, offshore wind turbines have a markedly higher energy yield than wind turbines onshore.

There are offshore wind turbines with a capacity of about 180 megawatts installed in Germany today. Up until August 2011, 23 offshore wind farms were authorized in the German Exclusive Economic Zone (EEZ) of the North Sea and an additional three in the Baltic Sea. The wind farms that are at the planning stage and under construction comprise a total of 1,851 turbines, which corresponds to a capacity of about 9,000 megawatts. The German government's target is an installed capacity of 25,000 megawatts by 2030. This would cover around 15 percent of electricity production in Germany.

However, the expansion of renewable energies is associated with interventions in nature. Mainly because of nature conservation concerns, the turbines are erected far away from the coast (30 to 100 km) and at water depths of 20 to about 50 m.

Only projects with not more than 80 turbines per project have been approved to date, the intention being to gather experience with the ecological impacts of offshore wind farms. These projects will form the basis for decision-making on a possible further expansion of offshore wind energy.

What impacts do these installations have on the marine environment? Let me just mention the most important ones:

- Electromagnetic fields around cables may affect the migratory behaviour of sharks, rays and some species of bony fish.
- Birds may collide with turbines, and a barrier effect and the loss of resting and feeding areas may ensue.
- Marine mammals (porpoises, grey seals and common seals) may be affected by the noise caused by pile-driving work for the foundations and as a result be displaced or physically injured.

The UBA is currently devising 'ecological guard-rails' and measures for an environmentally sound expansion of offshore wind energy which are designed to prevent and limit potential damage to the marine environment. One 'guard-rail' for protection of marine mammals are noise exposure limits which UBA recommends should be complied with during the erection of offshore wind turbines. Compliance with these limits requires the use of technical noise abatement measures such as bubble curtains or, better yet, alternative foundation laying techniques causing minimal noise. Overall, the negative 'side-effects' of offshore wind power also need to be evaluated in relation to the significant damage and risks which the use of conventional energy involves.

2.4. *Ocean Energy*

The harnessing of ocean energy through the 'might of water' is a use that is still at the research and development stage around the world, although the first technical innovations have recently been presented. Pilot projects such as the 300 kW prototype of the marine current turbine Seaflow-1, installed off the coast of England, the SeaGen marine current turbine currently being installed in Northern Ireland, and the world's first osmotic power plant, built in Tofte on the Oslo fiord, Norway, are testimony to the technical progress in this field. 'Ocean energy' denotes forms of energy like tidal, wave, current, salinity gradient (osmosis) and temperature gradient. These forms of energy have a rather small usable potential in Germany, according to current knowledge, but the global potential for using ocean energy is evaluated as being quite high.

Ecological 'guard-rails' to limit the potential impacts of these forms of energy have yet to be established. Here too, any potential negative impacts need to be evaluated in relation to the considerable damage and risks which the use of conventional energies involves.

2.5. *CCS as a Bridging Technology*

To mitigate climate change, we need a sustainable climate policy, aimed at increasing energy efficiency and using renewable energy sources. Storage of CO_2 in geological formations deep below the seafloor is currently under discussion as an interim technology. Possible sites for such storage could include formations of former oil and gas deposits in the North Sea, for example.

However, storing CO_2 under the seabed is problematic in several respects and is presumably more problematic than storage on land. The risk of CO_2 escaping to the water column, for example as a result of technical defects or accidents during transport, injection and storage or due to the use of unsuitable geological formations, is difficult to control. Large quantities of CO_2 could be released to the sea by leakage and change seawater pH and CO_2 concentrations. This would pose a threat in particular to those species which have calcium skeletons or shells such as certain algae, corals, snails and mussels. To ensure adequate safety during carbon storage and prevent potential impacts from carbon leakage, only justifiable rates of leakage should be allowed. We are therefore calling for—as is the German Advisory Council on Global Change—a limitation of leakage rates to less than 0.01 percent per year. This would limit allowable releases of stored carbon dioxide to less than 1 percent in 100 years.

In May 2011, the ECO2 project supported by the European Union was launched. Led by the Kiel-based Leibniz Institute for Marine Sciences (IFM-GEOMAR), a consortium of 27 institutes from nine countries will be studying the environmental impacts of sub-seabed CO_2 storage over the next four years. We are looking forward to the results.

2.6. *Marine Geo-Engineering*

I already mentioned at the beginning that the seas are our biggest and most sustainable carbon sink. It is therefore no surprise that there are geo-engineering or climate-engineering proposals which aim at manipulating this sink. Several methods have been proposed, of which the most prominent is ocean fertilization. It is also the only method which has been studied in experimental research. This method introduces micronutrients such as iron or macronutrients like nitrogen or phosphate to enhance algae production to thereby cause more organic material to sink and thus more carbon to be stored in the deep sea. Theoretical calculations of ocean fertilisation's potential initially raised much enthusiasm among the experts. If the efficiency of the biological pump could be increased by 10 percent, an additional 1 Gt of carbon could be removed from the atmosphere each year (about 12 percent of anthropogenic CO_2 emissions). However, the 13 iron fertilization trials carried out since 1993 brought sobering results. Although large algal blooms formed in all the experiments, only a few studies demonstrated a net export of carbon into deep waters. Proof of the effectiveness of ocean fertilization for the climate could not be produced to date. In addition, there has been mounting evidence of negative ecological impacts, for example toxic algal blooms, oxygen deficiency and the release of greenhouse gases (methane and nitrous oxide). Given that only intact marine ecosystems exhibit the resilience they need to withstand and mitigate the consequences of climate change, we should not jeopardize the health of these ecosystems. Therefore, rather than propagating risk-laden marine 'geo-engineering', we should better protect marine ecosystems. Germany must therefore continue to act as a driving force in the international climate negotiations and move forward quickly with national adaptation and reduction strategies.

2.7. *Change in Existing Uses*

In addition to the specific uses aimed at reducing atmospheric CO_2, we can expect changes in the type, intensity and extent of conventional uses to adapt them to the consequences of climate change and, as a result, changes in their impacts on the marine environment. This may lead to new conflicts between uses and between them and marine protection, and requires anticipatory regulation. Such adaptations concern near-shore uses such as coastal protection against rising sea levels as well as offshore activities such as a shifting of fishing areas, periods and types in response to emerging biological changes in the seas. Climate protection and marine protection therefore call for a holistic approach.

3. Implementation Needs and Successes

Any contribution to an international law conference should also deal with legal aspects. I will do so in the following, exploring the question of how the law is responding to the technical challenges. This concerns the legal responses with regard to, firstly, preserving the seas' carbon-sink function and resilience and, secondly, reducing the environmental impacts of marine uses.

It is clear that regulations on the use and protection of the seas exist at the international, European and German levels.

As far as preserving the carbon-sink function and resilience is concerned, in other words marine protection proper, the outcome is sobering. Effective protective instruments are not sufficiently in place so far. Whilst 18 percent of terrestrial area has been placed under protection, the corresponding figure for the world oceans is just around one percent. Although some provisions for protection of species and on protection areas exist, these measures have been insufficient so far. Some first successes to be mentioned are the designated marine protected areas under the Natura 2000 network and the designation of protected areas on the high seas agreed last year under the OSPAR Convention. As long as there are no binding management plans guaranteeing the protection of marine organisms and habitats, nothing has in effect been achieved for the marine environment as yet.

In addition to designating protected areas and implementing protective measures in a targeted way, it will be necessary in future to regulate uses of the seas which have a negative impact on marine life more effectively and to better coordinate them using spatial planning instruments in order to strengthen the resilience of the seas. This is particularly true for fisheries as one of the greatest pressures on marine ecosystems. According to the European Commission, the EU's common fisheries policy which regulates this sector has been one of fleet overcapacities and heavy subsidies, with the known effects of overfishing, low economic resilience and a decline in the volume of fish caught by European fishermen. The quotas proposed by the International Council for the Exploration of the Sea for use to set total allowable catches were raised regularly, and ecosystem protection requirements were broadly disregarded.

When large, predatory fish are exploited at high rates, representatives of lower levels of the food web have more scope to reproduce. For example, the exploitation of other large sealife feeding on jellyfish, such as turtles, is enhancing the development of medusas, whose abundance has increased greatly around the world not only because of this, but also for other reasons such as marine warming or eutrophication. This process is at a particularly advanced stage off the coast of Namibia. In the heavily overfished waters of the Benguela Current, jellyfish are now occurring in larger biomass volumes than do fish. Scientists believe that this is a profound change to the marine ecosystem which may have potential effects on the carbon cycle and the recovery of fish stocks. Jellyfish have few predators. In addition, they catch fish eggs, fish larvae and organisms fish feed on from the water, so that the current situation could possibly be irreversible.

Navigation, extraction of resources (as in the case of the Deepwater Horizon drilling rig), excessive nutrient inputs, and pollution of the seas by noxious substances, underwater noise and waste are further complex problem areas which require further legal standards and commitments.

As regards the regulation of climate engineering measures by international marine environmental law, there are essentially two messages I wish to convey: Firstly, international marine environmental law has achieved astonishing regulatory successes in this area. Some of these regulations address hitherto unregulated issues and are therefore of a paradigmatic nature. The second message is that despite these regulatory successes, international marine environmental law must be further developed also in relation to climate engineering.

Allow me to expand on the first point: The first international legal requirements on CCS were adopted under the London Convention and London Protocol. The Contracting Parties agreed, for environmental reasons, to allow CO_2 to be sequestered only under the seabed, not however in the water column. They further agreed that CO_2 streams for sequestration must consist overwhelmingly of carbon dioxide to prevent the sequestration of hazardous incidental components. Finally, technical requirements were adopted in so-called 'specific waste assessment guidelines'. I may say that the requirements of international marine environmental law in this area are standards which I consider to be necessary and appropriate from the environmental viewpoint. Significantly, these requirements of international marine environmental law had a key influence on the European Union's CCS Directive as well as the German carbon dioxide storage Act.

The second example is the regulation of ocean fertilisation under the London Convention and its Protocol. It was in the summer of 2007, in the beautiful town of Santiago de Compostela, that the Contracting Parties were confronted with the issue of ocean fertilization for the first time. Only one and a half years later, in London, they adopted a resolution which could point the way towards the regulation of numerous climate engineering measures. The resolution has a clear message: Ocean fertilization activities conducted on a commercial basis are banned and Contracting Parties should only allow scientific research projects. The other, equally important message: Even research projects can be problematic and must be monitored.

Furthermore, within a span of two years, the Parties to the London Convention developed an assessment guideline, the so-called Assessment Framework. This Framework provides specific criteria which now can and should be applied for assessing whether a project actually constitutes scientific research and whether it will have no adverse environmental impacts.

So, in the area of ocean fertilisation, international marine environmental law is leading the way in three respects: Firstly, research is allowed but application is not. Secondly, climate engineering research must be monitored. Thirdly, for the first time, standards for the assessment of research projects have been established in international law.

Admittedly, the Parties to the Convention on Biological Diversity decided, in 2008, to only allow small-scale research in the area of ocean fertilisation. The achievement of international marine environmental law is that it has fleshed out and operationalised this political agreement.

International marine environmental law, as well, is in need of reform and must face up more to the challenges of climate change. Of course, when the London Convention and the Protocol to the London Convention were adopted, nobody thought of CCS or ocean fertilization. That is why the Parties to the Convention and Protocol have been trying for three years, in various negotiating rounds, to find a sustainable, legally binding solution for marine climate engineering. The basic idea is that a separate, legally binding monitoring and assessment system be established for marine climate engineering. Of course, this raises a multitude of unresolved technical and legal questions and of course, opposition has to be overcome. The Federal Environment Agency has not found the philosopher's stone, either. The Parties are slowly coming closer to a solution. What is important, and on this there is consensus amongst the Parties, is that the substantive messages of the pilot resolution of 2008 should form the starting point for a binding regulation.

Why has international marine environmental law been so successful in the cases I described? For one thing, it is very helpful that the Contracting Parties agreed on a common objective, protection of the marine environment, and that this objective, however diverse the positions on details, forms a common basis for negotiations and decisions. For another, the negotiating situation under the LC/LP is relatively straightforward, because of the common objective to which all parties have committed and also because the negotiations essentially revolve around just one topic, protection of the marine environment, which makes for manageability and the possibility to bargain, to give and take. Finally, the successes are also due to the fact that the LC/LP have effective organizational structures. A Secretariat provides organizational assistance. In addition to the Conference of the Parties, a 'Scientific Group' meets, which ensures interdisciplinarity in working on problems and the integration of the necessary expertise.

Let me point out that these aspects also suggest that the regulation of climate engineering should not be entrusted solely to the Framework Convention on Climate Change. The Federal Environment Agency believes that the specific international regimes, such as the conventions on the protection of the seas for example, should in any case have a strong voice in this process.

So much for international law. Now, what contribution is European law making to solving the issues raised? Of prominent importance in this context is the Marine Strategy Framework Directive, which was adopted in the summer of 2008. However—let me mention this briefly—there are a number of other directives which are important to the protection of the seas, such as the Water Framework Directive, the Birds Directive and the Habitats Directive. Many further EU environmental regulations are also relevant here, since—as you probably know—pollutant inputs into the seas for example, mainly originate from land-based sources.

Coming back to the Marine Strategy Framework Directive, this Directive is the environmental pillar of the EU's maritime policy. It requires Member States to achieve good environmental status of their marine waters by 2020 in a multi-step planning process. An initial assessment must be prepared by mid-2012, monitoring programmes must be developed by mid-2014 and programmes of measures by the end of 2015. The Directive is based on an ecosystem approach, it strikes a balance between use and protection of marine waters. 'All clear' for Europe's seas? Unfortunately not. The MSFD has some basic weaknesses which make it questionable whether it may be expected to bring major improvements for the seas.

These weaknesses are, firstly, a lack of standards specifying the general objective 'good environmental status'. This means that the Member States themselves will have to define the desired level of protection. Secondly, the methodological standards for deriving specific targets are also insufficiently defined. As the seas have no natural boundaries, Member States will have to harmonise their standards. Due to the absence of guidance, this coordination will require considerable input. Thirdly, the Member States may allow exceptions to the requirement to achieve 'good environmental status' without any major legal hurdles.

So, the assessment is rather sobering. No specific provisions in regard to climate change, unclear objectives and methodological standards, more coordination work, broad exceptions—those are the drawbacks. On the positive side, much additional knowledge of the seas will be generated and made available to the public during the planning steps I mentioned. New issues and stress factors such as underwater noise or marine litter will be taken into account and evaluated for the first time. The resulting findings will certainly influence all areas of European maritime policy, so that the Directive can be expected to have positive effects for the protection of the marine environment at least in the medium and long term.

To conclude, a word also on the importance of our marine conventions HELCOM, for the Baltic Sea, and OSPAR, for the North-East Atlantic. Protection of the marine environment has been the focus of these conventions for several decades. During that time, they have developed ambitious approaches to the assessment and management of marine ecosystems in regard to many issues, such as 'marine eutrophication'. It is important to strengthen the protection efforts of the marine conventions and to also use these as a basis for the Marine Strategy Framework Directive. European marine environment policy would gain a lot from doing so. There is a growing understanding that many things can be regulated better at the regional level.

Also, the regional marine conventions can and should play an important role in coordination and harmonisation between the various transnational maritime regions.

4. Concluding Remark

This brings me to the conclusion of my address. I have tried to make it clear that even the adoption of the MSFD did not yet constitute a final victory for the protection of the seas. We will continue to need people dedicated to protecting the seas at public authorities, research institutes and NGOs to further analyse and identify problems and develop solutions to them. I remain hopeful that together we will be able to further improve the protection of the seas, even in times of climate change.

DISCUSSION SUMMARY
FORUM 1: PROTECTION OF THE MARINE ENVIRONMENT AND CLIMATE CHANGE

Lilly Weidemann[1]

1. Introduction

The forum on 'Protection of the Marine Environment and Climate Change' stressed the important role of the oceans in the global climate system, *inter alia* for the water and carbon cycle.

It also illustrated the manifold threats that are posed on the marine environment. The hazards firstly derive directly from the consequences of climate change and include rising water temperatures or ocean acidification. Secondly, they result from attempts to mitigate climate change, e.g. via carbon capture and storage (CCS). It was stressed that these perils add to the existing problems deriving from multiple uses such as shipping, fishing, resource extraction etc. and the implied effects on the marine environment like pollution, overexploitation and many more.

The resulting threats not only impair the ocean's capacity to mitigate climate change e.g. by the oceans' function as CO_2 sink, but also imperil marine biological diversity.

2. Discussed Topics

There was a broad consensus among forum participants that in light of the identified hazards the regulation of marine environmental protection is highly unsatisfactory.

2.1. *Hard or Soft Law Approach*

The discussion centred on the way forward to improve regulation of the marine environment and specifically on the question of whether a hard or a soft law approach is preferable.

---

[1] Lilly Weidemann, LL.B., Doctoral candidate (International Max Planck Research School for Maritime Affairs).

Although subject to considerable debate among legal scholars, hard and soft law are generally distinguished by their binding or nonbinding character.[2] Some scholars, however, conduct the characterisation along the three parameters precision of rules; obligation; and delegation to a third-party decision maker.[3] Accordingly, '[t]he realm of "soft law" begins once legal arrangements are weakened along one or more of [these three] dimensions'.[4]

The discussion showed that the differentiation between soft and hard law in the sense of the first definition is not a clear-cut issue in many cases. Jochen Flasbarth reported about the experience from the Convention on Biological Diversity (CBD),[5] where there are differing views within the different departments of the Federal Environment Agency (UBA) whether the decisions of the Conference of the Parties (COP) create legally binding obligations on the Member States.

Catherine Redgwell pointed out that the decisive quality of a legal instrument lies in its normative pull, which depends on a range of issues, e.g. its hortatory or mandatory wording, and not only on its binding or non-binding character.

2.2. *Fragmentation*

Another matter the discussion evolved around was the issue of fragmentation and the ensuing risk of conflicting obligations. Fragmentation of international law refers to the 'emergence of specialized and (relatively) autonomous rules or rule-complexes'[6] such as trade law, environmental law, law of the sea etc. The issue comes to the fore where parallel regulations applicable to the same matter trigger the question of which one should be applied to a given case.

---

[2] See e.g. Dinah Shelton (ed), *Commitment and Compliance: The Role of Non-Binding Norms in the International Legal System* (Oxford Univ. Press 2000).

[3] Kenneth W. Abbott and Duncan Snidal, 'Hard and Soft Law in International Governance' (2000) 54 International Organization 421.

[4] Ibid., 422.

[5] Convention on Biological Diversity, concluded 5 June 1992, entered into force 29 December 1993.

[6] International Law Commission, Fifty-eighth session, Geneva, 1 May–9 June and 3 July–11 August 2006, Fragmentation of International Law: Difficulties arising from the diversification and expansion of International Law, Report of the Study Group of the International Law Commission, finalized by Martti Koskenniemi, A/CN.4/L.682 (13 April 2006) 11.

In international environmental law, fragmentation is a common issue. In particular, since the first United Nations Conference on environmental issues in Stockholm in 1972,[7] the number of international environmental agreements on the global, regional and sub-regional level has greatly increased.[8] As a consequence, frictions among different environmental legal regimes occur. In the worst case, competitive regulations can impose mutually exclusive obligations on states.[9]

An example of possible conflicting obligations in the realm of the forum's topic concerns the obligation to fight climate change and to that end engage in geo-engineering and the competitive obligation to protect the marine environment and therefore refrain from geo-engineering. Redgwell argued that the first step to solve the problem should be to assess to which extent the applicable treaties address the problem of conflict of law. She also encouraged cooperation between the relevant Convention bodies to deal with the matter. Flasbarth drew attention to the importance of strict interpretation of the scope of the applicable agreements.

2.3. *Legal Basis for CCS on the (Extended) Continental Shelf*

In the context of mitigation of climate change by CCS, the forum furthermore discussed the legal basis for this activity on the (extended) continental shelf.

According to Article 77(1) of the United Nations Convention on the Law of the Sea (UNCLOS),[10] '[t]he coastal state exercises over the continental shelf sovereign rights to explore and exploit its natural resources'. Whether the exercise of subsoil carbon storage on the continental shelf is covered by this provision is subject to debate.[11]

---

[7] United Nations Conference on the Human Environment, Stockholm 5 to 16 June 1972, see Declaration of the United Nations Conference on the Human Environment, 21st plenary meeting, 16 June 1972, Chapter 11 <www.unep.org/Documents.Multilingual/Default.asp?DocumentID=97&ArticleID=1503&l=en> accessed 05 May 2012.

[8] Cathrin Zengerling, 'Sustainable Development and International (Environmental) Law—Integration vs. Fragmentation' [2010] Zeitschrift für Europäisches Umwelt- und Planungsrecht 175, 175.

[9] Gerhard Hafner, 'Pros and Cons Ensuing from Fragmentation of International Law' (2003–2004) 25 Michigan Journal of International Law 849, 851.

[10] United Nations Convention on the Law of the Sea, signed 10 December, entered into force 16 November 1994.

[11] See Alexander Proelss and Kerstin Güssow, 'Carbon Capture and Storage from the Perspective of International Law' in Christoph Herrmann and Jörg Philipp Terhechte (eds) *European Yearbook of International Economic Law 2011* (Springer 2011) 151.

This wording could be understood as restricting the use of the continental shelf to the exploration and exploitation of the natural resources as defined in Article 77(4) UNCLOS. On the other hand, Articles 80 and 60(1)(b) UNCLOS allow coastal states to construct installations for any economic purpose and to authorise and regulate the construction, operation and use of such installations.[12] This argues in favour of an exclusive competence of the coastal state for the regulation and conduct of CCS.

Redgwell emphasised that the exercise of CCS by the coastal state is particularly problematic on the extended continental shelf because of the overlying high seas column. According to Article 78(2) UNCLOS, the sovereign rights conferred to the coastal state by paragraph 1 of the provision 'must not infringe or result in any unjustifiable interference with navigation and other rights and freedoms of other States'. Thus, the coastal state has to consider the high seas freedoms of other states when exercising CCS on its extended continental shelf.

2.4. *International Efforts to Reduce Global Emissions*

Following Mojib Latif's presentation, the forum discussed the reasons for the stagnancy in international efforts to reduce global emissions. In this context, the role of the USA and China—the two states that together produce almost 50% of the global emissions—and the domestic political conditions for the commitment to greenhouse gas emissions reduction in the USA were reconsidered.

[12] Ibid., 156.

PART FIVE

OFFSHORE WIND ENERGY

WIND ENERGY AND MARINE ENVIRONMENT PROTECTION

Ursula Prall[1]

1. Introduction

1.1. *Expectations*

Electricity generated by wind turbines at sea is expected to substantially contribute to Europe's future and 'climate friendly' electricity supply. As of today, approximately 3.000 MW of offshore wind capacity are installed in Europe. The political objective, as reflected within the national action plans developed in accordance with the RES Directive,[2] comprises more than 40.000 MW to be installed by 2020.[3] Leading markets are to be expected in the following countries:[4]

- UK: 13 GW;
- Germany: 10 GW;
- Netherlands: 4 GW; and
- Spain and France: 3 GW each.

The advantage of offshore wind energy is that wind is both stronger and steadier at sea. For the North Sea, approximately 4.000 full load hours are expected, which means that electricity yields would be significantly higher than onshore and that fluctuations are less pronounced.

However, installing such a number of wind turbine generators (WTG) at sea (which would reach approximately 8.000 to 10.000, depending on their capacity) will foreseeably cause conflicts. Furthermore, 40.000 MW is not the final target. Germany alone, for example, aims to install 25.000 MW by 2025–2030.[5]

---

[1] Dr. Ursula Prall, Kuhbier Rechtsanwälte, Hamburg.

[2] Directive 2009/28/EC of the European Parliament and of the Council of 23 April 2009 on the promotion of the use of energy from renewable sources [2009] OJ L 140/16.

[3] Windenergieagentur Bremerhaven/Bremen, Offshore Windenergie 2011/2012, 21.

[4] Ibid.

[5] Strategy of the German Government on the Use of Off-Shore Wind Energy (2002) <http://www.bmu.de/files/pdfs/allgemein/application/pdf/offshore.pdf> accessed 05 July 2012, confirmed by Federal Government, Energiekonzept für eine umweltschonende,

1.2. *Resulting Conflicts*

Conflicts between offshore wind farms and other (legitimate) claims, interests and uses are predictable. This is because of the space required for offshore wind farms of the size and capacity which is needed in order to fulfil these expectations.

The most important other uses are probably shipping traffic and nature conservation. However, there are also other uses that could conflict with offshore wind farms, such as:

- fishing,
- military uses,
- pipelines and cables (data, communication, electricity, gas),
- oil and gas exploitation,
- sand and gravel exploitation, and
- air traffic.

These uses are only partly compatible with each other, which makes it necessary to reconcile them. Basically, this can be achieved by spatial arrangements, thereby separating conflicting uses as much as possible and/or necessary, and by reducing those of their effects which are detrimental to compatibility.

1.3. *In Particular: Approaches to Combine Nature Conservation and Offshore Wind in Germany*

In Germany both approaches are chosen as far as the use of offshore wind energy and nature conservation are concerned. The major practical issues which need to be tackled and eventually solved are:

- How are nature conservation issues dealt with during the permission procedure as well as during the construction and operation period?
- How and to what extent is habitat protection safeguarded?
- How and to what extent is species protection achieved?

---

zuverlässige und bezahlbare Energieversorgung (2010) <http://www.bundesregierung.de/Content/DE/_Anlagen/2012/02/energiekonzept-final.pdf?__blob=publicationFile> accessed 5 July 2012.

On the European level, the legal basis and the legal measure for these questions are notably the EIA,[6] the Habitat[7] and the Bird Directive.[8]

2. Nature Conservation in Consent Procedures

2.1. *Consent Procedure*

In Germany, wind farms are mandatorily subjected to an EIA if they consist of more than 20 wind turbines (Art. 4(1) and Annex II of the EIA Directive). Offshore wind farms generally fit within this mandatory provision; they typically consist of 80 WTG.

Conducting an EIA means that information needs to be provided which includes at least (Art. 5 (3) EIA-Directive):

- a description of the project comprising information on the site, design and size of the project,
- a description of the measures envisaged in order to avoid, reduce, and if possible, remedy significant adverse effects,
- the data required to identify and assess the main effects which the project is likely to have on the environment,
- an outline of the main alternatives studied by the developer and an indication of the main reasons for his choice, taking into account the environmental effects, and
- a non-technical summary of the information mentioned previously.

Before, scoping is conducted which means that the competent authority consults the developer about the information which needs to be supplied.

In practice, the competent authority which is the Federal Maritime and Hydrographic Agency (BSH) has provided some detailed guidance on what kind of information, and to what extent and acquired by which methods, is deemed to be necessary. This standardized concept is publicly

[6] Council Directive 1985/337/EEC of 27 June 1985 on the assessment of the effects of certain public and private projects on the environment [1983] OJ L 175/40 (as amended).

[7] Council Directive 1992/43/EEC of 21 May 1992 on the Conservation of natural habitats and of wild fauna and flora [1992] OJ L 206/7 (as amended).

[8] Directive 2009/147/EC of the European Parliament and of the Council of 30 November 2009 on the conservation of wild birds [2009] OJ L 20/7.

available.[9] This sometimes leads to the somewhat 'topsy-turvy' situation that developers have more or less finished gathering data when scoping takes place. Data acquisition/collection is required for two successive years, covering:

- avifauna, differentiating between migratory and sea- and waterbirds,
- marine mammals,
- fish, and
- benthos.

The actual EIA covers more issues, such as landscape, climate, sediment, biodiversity. In addition, the EIA needs to describe and evaluate in detail the possible impacts on the environment and mitigating possibilities. All information gathered during the EIA process must be taken into consideration in the consent procedure (Art. 8 EIA Directive.).

In Germany, consent for offshore wind farms has usually been granted except for two wind farms in the Baltic Sea which would have affected bird life too severely. If consent is granted, a number of collateral clauses are prescribed which need to be observed during the construction and operation periods of a wind farm.

2.2. *Construction and Operation; Decommissioning*

Those collateral clauses cover in particular:

- environmental monitoring during construction;
- environmental monitoring during operation for at least three years;
- reduction of noise emissions during construction from pile driving (see below);
- the possibility of turning off the turbines in cases of major bird migration events;
- an obligation of using only non-toxic materials; and
- an obligation of proper onshore waste disposal, including, obviously, that all spilling of hazardous substances is prohibited.

---

[9] BSH, Standard Investigation of the Impact of Offshore Wind Turbines on the Marine Environment (2007) <www.bsh.de> accessed 05 July 2012.

Also, those collateral clauses would deal with matters of safety for naval and air-borne traffic. This means that light emissions are caused which are an additional possibly detrimental effect on the environment.

Decommissioning is obligatory for turbines and foundation structures but this obligation might yet be subject to modifications in the future, as a consequence of possible repowering. In respect to cables it is contentious whether a removal would cause greater damage than leaving them in the seabed.

2.3. *Violation of Binding Provisions of Habitat and/or Bird Directives?*

In addition to an EIA which is fundamentally a matter of procedure and about information but not about binding material provisions, it is also—on the basis of this information—examined whether the project in question would lead to a violation of national nature conservation law, which, essentially, is governed by European law. This relates to habitat and species protection, as regulated by the Habitat and the Bird Directives.

3. Habitat Protection

3.1. *Designation of Protected Sites*

3.1.1. *Legal Obligation*

Both the Habitat and the Bird Directive oblige Member States to designate protected areas. Under Art. 4 of the Habitat Directive, sites need to be chosen and notified that either host natural habitats as listed in Annex I of said Directive or habitats of the species listed in Annex II. Under Art. 4 of the Bird Directive, protected areas are to be established which constitute special conservation measures concerning the habitat of Annex I species.

3.1.2. *Status in Germany*

In Germany, protective measures have been taken for slightly more than 30% of the Exclusive Economic Zone (EEZ). As of 2005, two special protection areas (SPA's) have been established under the Bird Directive, one in the North and one in the Baltic Sea.[10] Both areas are of importance as

[10] Verordnung über die Festsetzung des Naturschutzgebietes "Pommersche Bucht", BGBl. I 2005, 2778; Verordnung über die Festsetzung des Naturschutzgebietes "Östliche Deutsche Bucht", BGBl. I 2005, 2782.

breeding, moulting, and wintering areas as well as providing staging posts for migratory birds along their routes.

In addition, several sites of community interest (SCI's) were identified and notified to the European Commission in 2004. They have been included into the Commission's adopted lists of sites to be protected.[11] For those areas conservation status objectives have been prescribed.[12] According to Art. 4 (5) Habitat Directive, projects located within these sites are subject to Art. 6 (2), (3) and (4) of this Directive.

In respect to offshore wind energy, designating these SPA's and SCI's does not mean that those projects are categorically forbidden, since such an approach would collide with the International Law of the Seas. However, their impact on the site needs to be assessed in respect to their effects on the conservation status objectives. It is extremely likely that an offshore wind farm would have adverse effects on the integrity of the site concerned. That means that an exception would be needed, which is highly unlikely to be granted.

3.2. *Economical Instruments*

In Germany another instrument deterring project developers from planning a wind farm within a protected area has been established: As a rule a fixed, above market feed-in tariff has to be paid by the grid operator for electricity generated from renewable energy sources. However, the German Renewable Energies Act states that no such claim exists in respect to offshore wind farms which are located within a protected area. That means that the incentive system, which is after all the whole reason for developing the use of offshore wind energy by private parties, does not work. However, the provision applies only to wind farms which were licensed later than 2004, meaning that in one case a wind farm, which was licensed in 2002, still has a claim.

---

[11] Commission Dec. 2011/63/EU of 10 January 2011 adopting, pursuant to Dir. 92/43/EEC, a fourth updated list of sites of Community importance for the Atlantic biogeographical region, [2011] OJ L 33/52; Commission Dec. 2011/64/EU of 10 January 2011 adopting, pursuant to Dir. 92/43/EEC, a fourth updated list of sites of Community importance for the Continental biogeographical region [2011] OJ L 33/146.

[12] <www.habitatmare.de> accessed 05 July 2012.

3.3. *Evaluation*

With more than 30% of the EEZ protected, it can be said that at least an effective system of 'no-go areas' has been established, keeping areas which are considered to be the most valuable ones free from projects. No other wind farm has ever been applied for within a protected site—though quite a few are fairly close by. Indeed, it is conceivable from the map that the sites were shaped around existing projects which were in advanced proceedings instead of the other way round.

Although a lot of concern was raised from the offshore wind industry before these areas were designated/notified in 2005, altogether they have been accepted and definitely lead to more peaceful discussions as well as helped in respect to acceptance.

4. Species Protection

Species protection outside protected areas is more complicated and more contentious than habitat protection.

4.1. *Legal Obligation*

According to Art. 12 Habitat Directive, Member States are required to establish a system of strict protection of all Annex IV species in their natural range, prohibiting, inter alia:

- all forms of deliberate capture or killing of specimens of these species,
- deliberate disturbance of these species, particularly during the period of breeding, rearing, hibernation and migration, and
- deterioration or destruction of breeding or resting places of these species.

Under circumstances listed in Art. 16 Habitat Directive, exemptions—meaning: less than 50% of all cases—are possible.

Among those circumstances are imperative reasons of overriding public interest, including social or economic reasons and beneficial consequences of primary importance for the environment. Quite similar are species-related protective provisions of the Bird Directive. However, exemptions are more difficult to obtain under these provisions.

It is important to keep in mind the difference between:

- the prohibition on killing of specimens, which relates to individuals, and
- the prohibition on disturbance of species, which relates to populations.[13]

4.2. *Status of Species Protection in Germany*

4.2.1. *Marine Mammals (in Particular: Harbour Porpoises)*

Among the most relevant species concerned by offshore wind farms are marine mammals, and, in particular, harbour porpoises. At least within the German EEZ, hardly any other marine mammals occur. In particular, seals are usually found closer to the shore than the projected locations of the wind farms.

Harbour porpoises as well as seal species are Annex IV species (although not priority species) which means that they are subject to the strict protection regime established in Art. 12, 16 Habitat Directive. Studies conducted in the area of offshore wind farms already in operation show that harbour porpoises return to the construction site once construction has finished and are therefore not affected on a long term basis.

But construction noise caused by pile driving is a serious problem and may severely affect harbour porpoise individuals and even the population. However, the killing of harbour porpoises because of pile driving noise is extremely improbable since deterring methods are applied. That means that specimens are driven away from the immediate danger zone.

It is possible, though, that temporary or even permanent hearing threshold shifts occur which might lead to deterioration when foraging and mating and might possibly expose the animals to other dangers. For lack of better knowledge, it is momentarily assumed that no danger of hearing threshold shifts would occur if the noise is less than 160 dB (SEL) (Sound Exposure Level), which leads to an according collateral clause in the permission. The permission holder is obliged to take measures during construction to ensure that, outside a 750m radius, the noise impact is no more than 160 dB (SEL).

This sounds logical enough. Yet, the problem is that as yet no technology exists which promises such a noise emission reduction. Nonetheless, construction of offshore wind farms is already starting or under way, without applying noise mitigation systems reaching this limit.

---

[13] Guidance document on the strict protection of animal species of community interest under the Habitats Directive 92/43/EEC (2007) 35.

For the time being, two approaches are chosen. The animals are being deterred by seal scarers or pingers. Those devices are making noises which are 'uncomfortably' loud but do not constitute a danger so that animals are driven away and thereby kept out of harm's way.

On the other hand, the use of such devices poses a problem in itself. It is not yet fully discussed and evaluated but their use might contradict Art. 12 Habitat Directive. In my opinion it can be assumed that—should this be the case—their use is justified under Art. 16 (1) Habitat Directive and therefore in line with legal obligations: The devices are used to protect the animals from greater harm.

However, this is certainly not a long term perspective. More substantial reduction methods are necessary which lead to noise reduction at the source. All approaches conceivable at the moment are still in a research and development stage, such as drilling and so-called gravity foundations in deep water, or the use of hollow tubes. As promising as these technologies appear in respect to sound reduction, their applicability is not a given due to the following reasons:

- drilling might lead to problems with turbine stability,
- gravity foundations need certain geological sea bed conditions, and it is not yet tested whether this is a deep water (> 10m) option, and
- hollow tubes are huge structures which are a challenge to handle offshore, though probably next to gravity foundation the most promising option.

Bubble curtains are sometimes deemed to be state of the art. Only very recently, the necessary sound reduction has been achieved in some cases and it is hoped that further tests and optimisations will lead to a reliable technology. It should be noted, though, that their noise reduction potential does not necessarily guarantee compliance with the limit of 160 dB (SEL). This depends on the initial unmitigated noise level, which varies, depending, inter alia, on water depth, pile diameter and pile driving energy.

The next step would be to wait for a technology to be ready for use. But this would be a difficult step to take since it would mean that offshore wind development would need to be postponed for two or three years. Building an offshore wind farm is a long term business which needs decisions two or three years beforehand. A wind farm which starts construction in 2012 would have concluded the major contracts two years in advance at least. That would include logistics. Considering this lead time, it is simply impossible to integrate a new noise mitigation system or a break during mating and breeding phases of a few months at a late stage of project realisation which is already underway.

Very recently ambitious reduction systems have been tested. Some of the results were disappointing and the tested systems need further development, but some methods seem to be more promising. However, none of the noise reduction systems is serially applicable.

It has to be concluded that noise mitigation is a serious problem. From my perspective, as long as deterring methods and bubble curtains are used and research is continued, it can be tolerated for a pilot phase. However, it is legally impossible to build up the whole project pipeline by means of exemptions.

There is still a lot to be done; but due to strict handling by the competent authorities some technology forcing is now under way. It should be noted that noise emissions differ with every project. The necessary ramming energy differs with water depth, geological site conditions, and the foundation type. A standard solution will therefore not be found.

4.2.2. *Loons (gavia arctica and gavia stellata)*

In relation to birds, all European Bird species are protected under the Bird Directive, which means that the regime of Art. 5, 9 Bird Directive needs to be applied. A good solution was found for another problem which seemed to threaten offshore wind development in Germany until approximately two years ago. As to bird populations, which are identified in respect to biogeographical and not national boundaries, it has been the practice of the competent authority to allow a habitat reduction concerning 1% of the population. With about the twentieth offshore wind farm permission, 1% of the loon population (gavia arctica and gavia stellata) was affected by habitat loss—meaning that no other offshore wind farm would have been permissible if even one specimen (!) would have been displaced. Loons are a species particularly sensitive to vertical structures which keep a distance between 2 and 4 km from the outline of an offshore wind farm.

Another approach was necessary. The Federal Ministry of the Environment came up with the following: a main concentration area of loons was identified, and it was decided on the administrative level that no more projects than those already in 'legal existence' (i.e. permitted) should be given consent to within that area. That area covers more than the already designated SPA and is not a measure of habitat but of species protection. Outside this area, projects are still permissible (if no other reason for denying approval applies).

4.2.3. *Migratory Birds*

Finally an unsolved problem is protecting migrating birds from collisions with WTG. Obviously, an offshore wind farm needs to be marked during the night or during bad weather conditions. The exact effects of artificial lights on birds are still unknown. But experts agree that such lights might lead to collisions, exhaustion and/or disorientation, which usually ends fatally. Unfortunately, different species seem to react differently to certain colours which means that 'no green/no red/no blue' etc. is no answer. Another point experts agree upon is that the less light the better.

The consequent question is 'how much light is needed', and are there any differences between locations? The density of naval and in particular airborne traffic differs significantly within the German EEZ, and it is a fair question whether each wind farm would need the same amount of marking if little traffic is around. Not all authorities, until now, have been very open-minded about discussing this question.

4.3. *Evaluation*

Species protection is a matter which presents a couple of open questions—and it has the potential to seriously obstruct the development of the offshore wind industry. Since both nature conservation and the use of offshore wind industry are important concerns, it is necessary to continue searching for solutions. Those solutions might lie in the area of technological improvements as well as in reasonable concessions on a case-by-case basis.

5. Conclusion

For Germany it can be concluded that as to proceedings and habitat protection the legally necessary steps have been taken to ensure, as far as possible, a coexistence of the use of offshore wind energy and nature conservation.

In respect to species protection there are still important issued to be worked on. But it can also be said that during the previous years mutual acceptance and understanding have grown and that detrimental effects are likely to be not totally but at least partly solvable by technological improvements. Such improvements are under way and should be strongly pursued and supported.

WIND ENERGY AND MARITIME SPATIAL PLANNING

Monika Breuch-Moritz and Nico Nolte[1]

In 2002, the Federal government of Germany published a strategy for wind energy use at sea as part of its overall sustainability strategy. It is aimed at creating framework conditions allowing the potential for offshore wind energy to be exploited.

In December 2007, this was followed by the Federal Government's Energy and Climate Programme (IEKP), which formulates the goal of increasing the proportion of renewable energies in electricity production. The target for offshore wind energy is the provision of 25.000 MW by 2030. This goal was reaffirmed in the Energy Concept of the Federal Government from September 2010.

In implementation of this strategy, offshore wind farms have been licensed by the Federal Maritime and Hydrographic Agency (BSH). In the beginning this was done on a sectoral basis assessing only the effects on safety of shipping and on the marine environment.

In order to co-ordinate the growing conflict of maritime uses, in particular between developing and space intensive offshore wind farms and marine environmental protection goals as well as traditional maritime uses, an integrative and sustainable approach was needed for the development of the German Exclusive Economic Zone (EEZ).

The Federal Ministry of Transport, Building and Urban Affairs (BMVBS), as the competent ministry, set up targets (legally binding) and principles (guidelines that need to be particularly considered in the decision process) of spatial planning in the EEZ.

In this process ensuring safety and efficiency of maritime traffic as well as the protection of the marine environment and the freedoms (e.g. laying of cables and pipelines) guaranteed by the International Law of the Sea are of highest importance.

The German Spatial Planning Act (Raumordnungsgesetz—ROG), which has been made applicable to the EEZ in 2004, has since then provided the legal base for spatial planning in this sea area.

---

[1] Monika Breuch-Moritz, President and Dr. Nico Nolte, Head of Section Spatial Planning, Federal Maritime and Hydrographic Agency (BSH), Hamburg, Germany.

It is the goal of the Maritime Spatial Plan for the EEZ to determine co-ordinated regulations for following single uses and functions:

- shipping,
- exploitation of raw materials,
- pipelines and submarine cables,
- marine scientific research,
- energy production (especially wind energy),
- fisheries and mariculture,
- and the protection of the marine environment,

including regulations determining areas for single uses and functions. Further activities in the EEZ such as military uses, for which it is not possible on legal grounds to determine stand-alone regulations in a spatial plan for the EEZ, will find co-ordinated consideration within the regulations mentioned above.

The procedure for the determination of targets and principles of spatial planning for the German EEZ started in 2005, when hearings with agencies and non governmental organisations (NGOs) were held to discuss the scope of the accompanying environmental assessment report. In 2008, the BSH finished the draft of a spatial plan, including justifications, and an environmental report with the consent of the BMVBS. The environmental report according to the Strategic Environmental Assessment Directive describes and evaluates especially the likely significant effects of the implementation of the maritime spatial plan on the marine environment. These documents have been made available to the public and have been sent to agencies and NGOs for their comments. The neighbouring states have been informed; meetings with the Ministries responsible for spatial planning in these neighbouring states have taken place. Two public hearings have been held, one for the Baltic Sea (30 September 2008) and one for the North Sea (6 October 2008), at which the submitted comments have been discussed with more than 100 representatives from agencies and NGOs. After having evaluated the comments and discussions in the public participation phase, the draft has been amended. Because of the changes another phase of public and transboundary participation was carried out in spring 2009.

The basic structure of the spatial plan follows the analysis of the ship traffic based on AIS-information provided by the Water- and Shipping Administration. Because shipping has special weight according to UNCLOS, the main shipping routes are designated as priority areas, which must be kept free from obstacles. This designation is the result of Art. 60 para 7 UNCLOS, which rules that installations may not be established by the coastal State where they are capable of interfering with the use of recognized sea lanes essential to international navigation.

The second important goal is the designation of priority areas for offshore wind energy, which do not conflict with other uses and the marine environment. These areas plus the already licensed wind farms safeguard the realisation of roughly 12.000 MW, which is a major step toward delivering the ambitious goals for offshore wind.

The third important goal is the protection of the marine environment, therefore according to the spatial plan no wind turbines are allowed in Natura 2000 areas. As a basis for description and assessment of any substantial impacts on the marine environment that are likely to be caused by the implementation of the plan, a large scale Strategic Environmental Assessment (SEA) has been carried out—the first time in a sea area distant from the coast. The SEA showed that the implementation of the plan has had no significant effect on the marine environment.

The legal ordinance for the maritime spatial plan has been set into force by the BMVBS for the EEZ in the North Sea on 26 September 2009 and for the EEZ of the Baltic Sea on 19 December 2009.

Transnational uses and marine conservation issues make it necessary to cooperate effectively with the neighbouring countries and countries adjacent to the North Sea and Baltic Sea. Therefore, one of the future challenges is to intensify transboundary cooperation concerning inter alia designation of wind parks as well as energy grids, safety of shipping and the protection of the environment. The INTERREG Project 'BaltSeaPlan' which deals with the introduction of Maritime Spatial Planning in the Baltic Sea[2] is a good example for this approach.

2 <www.baltseaplan.eu> accessed 05 July 2012.

DISCUSSION SUMMARY
FORUM ON OFFSHORE WIND ENERGY: ACTORS, LEGAL INSTRUMENTS AND DECISION-MAKING PROCEDURES

Ronán Long[1]

1. Introduction

Many countries around the world are implementing programmes to reduce emissions of carbon dioxide and other greenhouse gases in line with the targets set down by the Kyoto Protocol to the United Nations Framework Convention on Climate Change (UNFCC).[2] In parallel with these initiatives, the European Union (EU) has steadfastly sought to ensure energy security, increase competition in the market place and endeavoured to make the energy sector more compatible with environmental objectives.[3] In recent years, it has actively sought to promote energy from renewable sources, which is given a very expansive definition in EU law and means: 'energy from renewable non-fossil sources, namely wind, solar, aerothermal, geothermal, hydrothermal and ocean energy, hydropower, biomass, landfill gas, sewage treatment plant gas and biogases'.[4]

Although offshore renewable sources such as electricity generated from wind turbines clearly comes within the scope of this definition, this source has traditionally only played a very minor part in the European energy market.[5] Nevertheless, the importance of this source is undergoing fundamental change as a result of the EU's evolving policy on climate change and as a consequence of the coming into force of the Treaty on the Functioning of the EU, which expressly provides for the first time in

---

[1] Dr. Ronán Long, Jean Monnet Chair of European Law, School of Law, National University of Ireland Galway, Ireland.

[2] UNFCC adopted 9.05.1992 and into force 21.03.1994. Kyoto Protocol adopted 11.12.1997 and into force 16.02.2005.

[3] Communication from the Commission, 'An overall view of energy policy and actions' COM (97) 167 final of 23 April 1997.

[4] Council Directive 2009/28/EC on the promotion of the use of energy from renewable sources and amending and subsequently repealing Directives 2001/77/EC and 2003/30/EC [2009] OJ L 140/16.

[5] Ludwig Kramer, *EU Environmental Law* (7th Ed., Sweet and Maxwell 2011) 380 et seq.

EU Treaties that one of the principal objectives of the EU energy policy is to promote the development of new and renewable forms of energy.[6] In pursuing this objective, regard must be had to the need to preserve and improve the environment.[7] Moreover, the Treaty goes on to provide that regulatory measures giving effect to the EU's energy policy 'shall not affect a Member State's right to determine the conditions for exploiting its energy resources, its choice between different energy sources and the general structure of its energy supply'.[8]

In this broader EU context, one of the jewels in the crown of first Hamburg International Environmental Law Conference 2011 is the Forum on Offshore Wind Energy. The importance of this session must be considered in light of a number of factors including: the findings of the International Energy Agency that renewable energy, and offshore renewable energy sources in particular, can make a significant contribution to the generation of electricity from carbon free sources and thereby facilitate the implementation of climate change mitigation and energy security policies at national, regional and global levels.[9] Indeed, it is unsurprising to note that the implementation of Germany's incredibly ambitious offshore renewable energy programme is being followed closely by the international community as concerns are raised about the failure of the Kyoto Protocol to deliver the anticipated reductions on anthropogenic impacts on climate change.[10] Moreover, the climate change mitigation strategy adopted by Germany, must now be viewed through the prism of the position taken by the EU at the 17th Conference of the Parties (COP 17) to the UNFCC in Durban regarding a second emissions commitment period (following on from the first period 2008–2012), as well as the planned negotiation of individual emissions targets at COP 18 in Qatar in 2013.[11] As such, it demonstrates a political willingness by a core EU Member State to shoulder some of the burden associated with the discharge of the principle of common but differentiated responsibilities for countries that are at different

[6] Art. 194 (1) (c) of the Treaty on the Functioning of the European Union.

[7] Art. 194 (1) of the Treaty on the Functioning of the European Union.

[8] Art. 194 (2), para 2, of the Treaty on the Functioning of the European Union.

[9] International Energy Agency, 'How the world energy sector can deliver on a climate agreement in Copenhagen' (OECD/IEA, 2009) 9.

[10] See, inter alia: Joseph E. Aldy and Rober N. Stavins (eds), *Post Kyoto international climate policy: Implementing architectures for agreement*, (Cambridge University Press 2010).

[11] J. Jackson Ewing, 'From Kyoto to Durban: The fits and starts of global climate change negotiations', *NTS Insight*, February 2012, Singapore: RSIS Centre for Non-Traditional Security (NTS) Studies.

stages of economic development.[12] Furthermore, it also appears to be one of the few instances where a developed State is adopting bold mitigation measures such as those that are frequently advocated by climate change scientists. The need for international leadership by an EU Member State on the world stage is all the more apparent and pressing as several major developed countries including the United States, Japan, Canada and Russia show little commitment to continuing the Kyoto Protocol process at a multilateral level.[13]

Against this background, it is indeed a privilege to report on the cut and thrust of the debate that took place at the Forum, where many of the vexing issues were discussed by a knowledgeable audience who had a first class understanding of various law and policy issues that permeate any discussion of the subject at hand. As a starting point, there appeared to be general acceptance by the Forum that offshore renewable energy ought to form a central strand in any programme aimed at mitigating the effects of climate change and boosting energy security. The focus of the discussion at the Forum centred around the three excellent inter-disciplinary papers delivered by the panel members, namely: Dr Ursula Prall who as Chief Executive Officer represents the Forum Winderenergie (this is representative body for wind energy developers and their industry partners); Monika Breuch-Moritz who is President of the German Maritime and Hydrographic Agency and an absolute authority on the subject matter of her paper—maritime spatial planning; and Gert-Jürgen Scholz who is a Director at the Federal Ministry of Transport, Building and Urban Development and in this capacity holds an onerous portfolio of responsibilities dealing with a very broad range of operational issues including maritime safety.

Much of the Forum's deliberations focused on the role played by the various institutional actors, legal instruments and decision-making procedures in authorising and regulating offshore wind energy development. More specifically, the three speakers addressed in a non-adversarial and informative style, the pressing topics of offshore energy development and its relationship with marine environmental protection, maritime spatial planning, as well as the safety implication for navigation in general and shipping activity in particular. Although the three papers dwelt on offshore renewable energy development in sea areas under the sovereignty

[12] Art. 3 of the UNFCC, which does not set down the obligation as a principle. See inter alia: Lavanya Rajamani, 'The Principle of Common but Differentiated Responsibility and the Balance of Commitments under the Climate Regime' (2000) 9 *RECIEL* 120.

[13] See: *Financial Times* (11 November 2011); *Guardian* (12 December 2012).

and jurisdiction of Germany (that is to say, in the German territorial sea and Exclusive Economic Zone, EEZ), this proved to be a major strength in the Forum as it provided a very useful paradigm which allowed the participants to debate the practical aspects of how best to resolve the various challenges that arise due to the large-scale wind energy development that it is planned for the German maritime area. The principal arguments and issues raised at the Forum are summarised here.

2. Offshore Wind Energy Development and the Obligation to Protect the Marine Environment

There was general agreement among Forum participants that there were significant advantages associated with offshore wind energy development in the German EEZ. These advantages, however, needed to be balanced very carefully against the conflicts that would arise in relation to the use of the marine environment for other well-established maritime activities such as shipping, commercial sea fishing activity and marine scientific research. What is more, a number of participants emphasised that the environmental impacts of offshore wind energy development needed to be reconciled with the various nature conservation obligations that arise under a broad range of national, European and international marine environmental protection instruments. There also appeared to be general agreement that this reconciliation of competing interests would not be an easy task and posed a number of immediate challenges for the Federal authorities.

Very helpfully, Dr Prall's paper shed considerable light on these matters by focusing on three key issues, namely: how environmental considerations are taken into account during the consent process authorising wind farm development in the marine environment; and the related matters of habitat protection and species protection under the Habitats and Birds Directives.[14]

Dr Prall's paper led to considerable discussion in the Forum as it appeared that the consent process for offshore wind farms in Germany

[14] Council Directive 92/43/EEC of 21 May 1992on the conservation of natural habitats and of wild fauna and flora [1992] OJ L 206/7; Council Directive 79/409/EEC of 2 April 1979 on the conservation of wild birds, [1979] OJ L 103/1. Subsequently codified by Directive 2009/147/EC of the European Parliament and of the Council of 30 November 2009 on the conservation of wild birds [2010] OJ L 20/7.

was relatively rigorous in so far as it required environmental impact assessment for all wind farms the were made-up up of over more than 20 turbines. There was some disagreement as to whether this was a high or low threshold. This regulatory requirement, nonetheless, accords with both the spirit and letter of the Environmental Impact Assessment Directive and the various transposition measures adopted by Germany.[15] Similarly, the requirement to collect data over a two year period for the environmental impact statement also appeared to be relatively onerous in its temporal limits and places a considerable scientific and financial burden on wind energy developers. A number of participants noted that consent has been granted in nearly all cases for offshore renewable energy development in Germany, albeit subject to conditions addressing many technical and environmental mitigation matters such as procedures for the decommissioning and removal of structures at the end of the operational life of the farms. There was some discussion and clarification regarding the refusal of consent for farms in the Baltic Sea due to their potential impact on the spatial areas that were scheduled to be designated as special protected areas under the Birds Directive.

This discussion led directly to the remaining issues addressed in the speaker's paper which pertained to the habitat and species protection under the principal European instruments, the Habitats and Birds Directives. All participants were again surprised to learn that almost 30% of Germany's EEZ has been designated as protected areas under these Directives. In terms of spatial scale, this appears to be an impressive figure at a comparative level.[16] Take, for instance, the entire EU *NATURA 2000* network, which is made-up of 159 designated sites covering 147.322 km^2 and amounts to only 1% of the entire OSPAR Maritime Area in the North-East Atlantic.[17]

In the German EEZ, although projects are not categorically prohibited in areas that are designated under the provisions of the Directives, consent may only be granted as an exception to the conservation objectives of the European nature conservation instruments. This restriction is further bolstered by the withdrawal of any entitlement to fixed feed-in tariff

---

[15] Directive 2011/92/EU of the European Parliament and of the Council of 13 December 2011 on the assessment of the effects of certain public and private projects on the environment [2012] OJ L 26/1.

[16] For comparative data, see, OSPAR Quality Status Report 2010 <www.ospar.org> accessed 05 July 2012.

[17] Ibid., 92.

for projects that were licensed later than 2004 in protected areas. In practice, Dr Prall informed the Forum that this has resulted in an effective system of no-go areas for development which amounts to 30% of the EEZ. Moreover, this requirement has ensured that most valuable sites from a nature conservation and marine scientific perspective are now protected in the German EEZ.

In view of the scale of renewable energy development in the EEZ, the participants were very interested in species protection outside of the aforementioned protected area. The discussion focused on the strict scheme of protection set down for Annex IV species in their natural range under Article 12 of the Directive. Two species were mentioned in Dr Prall's paper to illustrate the scheme of protection afforded by the Directive, harbour porpoises and loons which are a group of aquatic bird species and one of the oldest bird families found in the North Sea. The first issue addressed by the speaker, which is the subject of considerable controversial debate at an international level,[18] relates to the noise emissions associated with the installation of turbines in the seabed. As pointed out by the speaker, it is not technically possible currently to reduce emissions to the prescribed thresholds of 160 DB SEL (sound exposure level). There appears to be some agreement in the scientific literature that the noise impact on marine mammals is 'more severe during the construction of wind farms than during their operation'.[19] Thus it is understandable that present efforts are focused on deterring porpoises from remaining in the vicinity of the turbines during the period of construction of new wind energy installations. Although a number of scientific studies are said to establish that the porpoises return to areas where turbines have been constructed, some doubt was expressed by Forum members regarding the probity and utility of present measures aimed at managing the impacts of noise, particularly that the cumulative impacts of noise are poorly understood as noted by one participant. There was unanimous agreement that there is an urgent need for further research to address the impacts of noise emissions with a view to adopting appropriate and effective mitigation strategies. Moreover, a number of participants suggested that this issue is best addressed at a European level within the broader context of the

[18] P. Madsen et al., 'Wind turbine underwater noise and marine mammals: implications of current knowledge and data needs' (2006) 309 Marine Ecology Progress Series 279.

[19] Ibid.

implementation of the Marine Strategy Framework Directive.[20] On the subject of the loons, participants were again interested to learn that a 'concentration area' larger in size than the designated SPAs areas had been identified and that an administrative agreement had been agreed with developers to ensure that no projects would be approved within this extended area. Again a degree of surprise was expressed by Forum participants that some technical matters such as the protection of migratory birds from collision with turbines as a result of intensity of the light emitted from the artificial safety of navigation lights on the turbines had not yet been resolved at a technical level.

On the wider issue of protection, there was considerable discussion regarding the use of the derogation procedure under Article 16 of the Habitats Directive and the various thresholds that were been applied in relation to offshore renewable energy developments in the German offshore area. Similarly, a number of questions were raised regarding the legal basis in the 1982 United Nations Convention on the Law of the Sea (hereinafter, the '1982 Convention') which permitted the extensive development of the EEZ for offshore energy development and more specifically the damage that would be done to the continental shelf by the placement of 8,000 wind turbines in the seabed. Some participants expressed concern regarding the 'due regard' provision in Article 56 which required Germany to have regard to the rights and duties of other States and to act in a manner compatible with the provisions of the 1982 Convention. This issue is examined further on below.[21]

Overall, the Forum concluded that Germany was working steadily towards achieving a degree of coexistence between the nature conservation objectives that needed to be achieved under a range of European instruments and the strategic requirement to undertake intensive and large-scale renewable energy development in the German EEZ. Although

---

20 Directive 2008/56/EC of the European Parliament and of the Council of 17 June 2008 establishing a framework for community action in the field of marine environmental policy (Marine Strategy Framework Directive) [2008] OJ L164/19. For commentary on this Directive see Ronán Long, 'The EU Marine Strategy Framework Directive: A New European Approach to the Regulation of the Marine Environment, Marine Natural Resources and Marine Ecological Services', (2011) 29 Journal of Energy and Natural Resources Law 1; Arie Trouwborst, Harm M Dotinga, 'Comparing European Instruments for Marine Nature Conservation: The OSPAR Convention, the Bern Convention, the Birds and Habitats Directives, and the Added Value of the Marine Strategy Framework Directive' (2011) 20 *European Energy and Environmental Law Review* 129.

21 Para E infra.

several scientific and technical matters remained unresolved, the Forum participants acknowledged the degree of commitment shown by the relevant public and private parties engaged in offshore wind energy development to achieving appropriate technical solutions to the remaining problems.

3. Marine Spatial Planning and Wind Energy in the Germany EEZ

One essential feature in delivering the offshore renewable targets set by Germany is the relationship between the wind energy developments and maritime spatial planning (MSP).[22] The paper presented by Monika Breuch-Moritz sets out the policy background stemming from the publication of strategy for wind energy use at sea as part of the overall sustainability strategy in 2002, and from the various federal programmes that have been implemented in the intervening years. One of the guiding principles underpinning the policy which has much public support is the commitment not to undertake turbine construction in maritime areas where they are visible from the islands and the coast. As a general rule pertaining to licensing, offshore wind farms are assessed on the basis of impacts on safety of shipping and their impact on the marine environment. One of the objectives of the federal policy, which was much discussed in the Forum, is to resolve conflict as early as possible and to ensure that an integrated approach to decision-making is implemented as early as possible in the licensing process. In parallel with the rolling-out of the policy, the Ministry of Transport set up legally binding targets and a number of principles regarding MSP in the EEZ. The legal basis for MSP is in the German Spatial Planning Act 2004, which has specific provisions providing for a range of matters such as shipping, marine scientific research, pipelines, and the protection of the marine environment. Setting an example, for other Member States of the EU perhaps, is the fact that the determination of targets for MSP in the EEZ was subject to extensive public consultation before the draft spatial plan was adopted in 2008. The plan which was

[22] See inter alia: European Commission, 'Maritime Spatial Planning in the EU—Achievements and Future Development' COM (2010) 771 final; 'Roadmap on Maritime Spatial Planning: Achieving Common Principles in the EU' COM (2008) 791 final; Frank Maes, 'The international legal framework for marine spatial planning' (2008) 32 Marine Policy 797; Paul M. Gilliland and Dan Laffoley, 'Key elements and steps in the process of developing ecosystem-based marine spatial planning' (2008) 32 *Marine Policy* 787.

the outcome of this process was then assessed under the SEA Directive and a further round of consultations was undertaken with close to one hundred representative organisations that are concerned with offshore developments in the German EEZ.[23] As an aside, the moderator noted that this process accords fully with the general scheme of the obligations that arise under the United Nations Economic Commission for Europe Convention on Access to Information, Public Participation in Decision-making and Access to Justice in Environmental Matters, which is also part and parcel of EU law.[24] A number of Forum participants drew attention to the fact that the first priority in the basic structure of the plan is to allow for the designation of shipping areas as priority areas taking into account Art. 60 (7) of UNCLOS. The second objective is the designation of priority areas for offshore wind farms and the third important objective is the protection of the environment.

There was some discussion among participants regarding the hierarchy that pertains to these objectives and which objective ought to take priority in cases of conflict regarding a particular use. Interestingly, one of the reported outcomes of the assessment process undertaken in relation to the proposed developments in the German EEZ is that the SEA revealed that there was no significant impact on the marine environment that would be caused by the implementation of the plan. Indeed, Breuch-Moritz was at pains to emphasise that the maritime spatial plan is in force and is a practical instrument that facilitates the avoidance and resolution of conflicts between the various users of the marine environment. Several questions were posed regarding trans-boundary cooperation and the implementation of the ecosystem-based marine management at a regional level. Quite

---

23 Council Directive 2001/42/EC of the European Parliament and of the Council of 27 June 2001 on the assessment of the effects of certain plans and programmes on the environment [2001] OJ L197/30.

24 Council Decision of 17 February 2005 on the conclusion, on behalf of the European Community, of the Convention on access to information, public participation in decision-making and access to justice in environmental matters [2005] OJ L 124/1; Council Directive 2003/4/EC of the European Parliament and of the Council of 28 January 2003 on public access to environmental information and repealing Council Directive 90/313/EEC [2003] OJ L 4126; Council Directive 2003/35/EC of the European Parliament and of the Council of 26 May 2003 providing for public participation in respect of the drawing up of certain plans and programmes relating to the environment and amending with regard to public participation and access to justice Council Directives 85/337/EEC and 96/61/EC [2003] OJ L 156/17; Regulation (EC) N° 1367/2006 of the European Parliament and of the Council on the application of the provisions of the Aarhus Convention on Access to Information, Public Participation in Decision-making and Access to Justice in Environmental Matters to Community institutions and bodies [2006] OJ L 264/13.

clearly, there are a number of express requirements for transboundary cooperation under the Marine Strategy Framework Directive and the EU's Integrated Maritime Policy.[25] Attention was drawn to a number of EU and regional initiatives including the aptly named EU project: BalticSeaPlan.[26] Indeed, a notable feature of MSP has been coordination and cooperation by Germany with neighbouring States in both the Baltic and North Seas.

4. Offshore Wind Energy and Shipping

Clearly, one of the major impacts of offshore wind energy development is the impact on maritime navigation in general and shipping in particular. According to the information in the paper presented by Gert-Jürgen Scholz there are some estimates which indicate that up to 15% of the entire German EEZ will be covered by wind turbines once the various installation programmes are complete. Scholz explained that the Directorate General Waterways and Shipping is mandated with reviewing the safety aspect of proposed projects on shipping and the projects are assessed using a comprehensive set of guidelines and management tools that includes: factors such as the safety of shipping lanes; risk analysis (one collision per 100 years); block group of turbines; as well as requirements for marking, design standards, and proof of self-protecting elements around each turbine.

One related issue which was of major interest to those in attendance is the attribution of responsibilities of the various parties involved in the development process including the responsibility of the State for maritime safety and the operator's duty of self-protection against maritime accidents. The speaker gave the concrete example of the responsibility to provide tug coverage to ensure achievement of an acceptable collision rate of 1 collision per 100 years. Most participants were of the view that the cost and responsibility for the provision of this service ought to rest with offshore developers and with the Federal authorities. In addition, a number of contributors from the floor drew attention to fact that there were no legal impediments to the imposition of additional safety measures on ship operators. Many speakers drew attention to fact that it is

---

25 European Commission, 'An Integrated Maritime Policy for the European Union', COM (2007) 575 final of 10 October 2007 and Commission Staff Working Document, SEC (2007) 1278 of 10 October 2007. This followed the adoption of the Green Paper on a Future Maritime Policy for the European Union by the Commission on 7 June 2006.

26 <www.balticseaplan.eu>.

legally possibly to impose costs on the operators relating to safety and marine environmental protection services and this aspect of the legislation is under review in Germany.

Another issue which provoked considerable debate is the certification and inspection of the vessels that undertake the construction and maintenance of offshore installations such as wind turbines. Essentially, it appears that different standards apply to the shipping industry in different North Sea states. A number of specialists at the Forum argued that there appears to be a strong case for further regulatory harmonisation in this regard at an EU level and that this ought to be reviewed by the European Commission with a view to bringing forward draft legislation to address this matter.

The third issue is emergency response planning and operations. Here again the question of the responsibilities of the State and the operator's responsibilities are juxtaposed and many participants felt that the costs associated with the provision of expensive services such as helicopter coverage should not be borne by the Federal Government.

From an administrative law perspective, the difficulties arising from the administrative division of various functions was highlighted including the importance of assigning a number of core functions to dedicated agencies such as the Central Command for Maritime Emergencies Germany, which is the central body tasked with delivering services in relation to maritime emergencies and accidents. For most participants this appeared to be a prudent approach.

Again a central theme in the paper is coexistence of a maritime partnership between shipping and wind energy. The envisaged setting up of flight simulator in Cuxhaven is a positive development in this respect and the overall impression presented from the discussion arising out of Scholz's presentation is that offshore renewable energy development would not impede the safety of shipping to unacceptable level and thus thwart one of the fundamental objectives of UNCLOS and that is the global community's interest in the freedom of navigation. Indeed, it should not be forgotten that the economic rights afforded to Germany in the EEZ are without prejudice the legitimate rights and uses of all States in this zone.

5. General Remarks

Perhaps it may be appropriate to conclude this report by highlighting a number of common themes that were evident throughout the Forum deliberations. Some participants (including the moderator) expressed surprise at the scale of the proposed development which entails the construction of 8,000 offshore turbines in a relatively small spatial area of the North Sea and Baltic Sea. Several participants asked whether this level of intensive development was absolutely necessary. In reply, a number of contributors expressed the view that there were few alternatives to this approach as Germany's economy is highly energy intensive and per capita carbon dioxide emissions from the burning of fossil fuels is comparatively high at European and global levels.[27]

There was a much broader policy backdrop to the debate in so far as the shift away from nuclear energy production under the Federal Government's Energy and Climate Change Programme is profoundly influencing the shape and direction of climate change law and policy at federal and regional levels. In this regard, a sense of urgency in relation to offshore renewable energy development comes from the German Government's well publicised commitment to specific targets such as the requirement to generate 35% of Germany's future electricity needs from renewable sources by 2020, and to an incredibly ambitious 80% by 2050. Similarly, the objective of cutting the national electrical consumption to half of the levels consumed in 2008 by 2050 sets down incredibly ambitious targets, which will be followed closely by the academic community both within and outside of Germany. Essentially, many participants expressed the view that wind energy is the only technical feasible solution to achieving the objective of attaining the production of 25,000 MW from offshore renewable sources by 2030.

In view of the ambitious and fixed nature of the targets and the broader policy backdrop, it was interesting to see how quickly the Forum's deliberations moved from an academic discussion of the theoretical aspects of climate change into the *realpolitik* world of practical implementation of the various measures that are applicable to renewable energy production in the marine environment. There was general agreement that the framework for addressing these challenges is the one provided for by the United Nations Convention on the Law of the Sea. Suffice to note here that the 1982 Convention explicitly provides that the coastal State has sovereign rights for the economic exploitation and exploration of the zone and specifically mentions the production of energy from the

[27] This is borne out by the World Bank data, see: World Bank, CO2 Emissions (Metric Tons per Capita) <http://data.worldbank.org/indicator/EN.ATM.CO2E.PC> accessed 05 July 2012.

water, currents and the winds.[28] The exclusive rights of the coastal State extend to the construction, authorisation and regulation of installations and structures necessary for the exercise of such sovereign rights in the EEZ.[29] One participant drew attention to the concept of reasonableness in international law and the need to strike a balance between the navigation rights under UNCLOS and the right of coastal States such as Germany to undertake economic development within the EEZ. International law is relatively well settled on this topic and the coastal State is free to place turbines in the territorial sea and the EEZ as long as they are not an unreasonably interference with the right of innocent passage and the freedom of navigation and other rights as set out in the Convention.[30] In many instances this will require the designating sea lanes and prescribing traffic separation schemes. In such instances, the coastal State must take into account a number of factors including: the recommendations of the IMO, any channels customarily used for international navigation, the special characteristics of particular ships and channels, and the density of traffic.[31] In addition, due publicity must be given to sea lanes and traffic separation schemes on navigation charts.[32] Indeed, one commentary has suggested that recent practice regarding the practical aspects of designation of appropriate traffic separation schemes in the vicinity of offshore renewable energy development support the view that existing IMO mechanisms are adequate for managing shipping traffic and thus facilitate the preservation of navigational safety and the protection of the marine environment.[33]

One penultimate point relates to the interdependence of different economic sectors operating in the marine environment. In general, the Forum supported proactive measures adopted by Germany to promote offshore renewable energy development and the various steps that are been taken to reconcile the programme with the various rights and

---

[28] Art. 56 (1) (a) of UNCLOS.
[29] Art. 60 (1) of UNCLOS.
[30] Arts 17–26, 56 (2), 58 of UNCLOS.
[31] Art. 22 (3) of UNCLOS.
[32] Art. 22 (4) of UNCLOS.
[33] David Leary and Miguel Esteban, 'Climate Change and Renewable Energy from the Ocean and Tides: Calming the Sea of Regulatory Uncertainty' (2009) 24 The International Journal of Marine and Coastal Law 617, who give the example of the International Maritime Organization, Sub-Committee on Safety of Navigation, Routing of Ships, Ship Reporting and related Matters. Amendments to the Traffic Separation Scheme 'Off Lands End, Between Longships and Seven Stones' IMO Doc NAV 54/3/5 of 28 March 2008 <www.sjofartsverket.se/pages/14913/54-3-5.pdf> accessed 05 July 2012.

duties that arise under UNCLOS. Again, however, one weakness in the presentations was the absence of an overview of the various measures that have been taken at a multilateral level to provide an appropriate regulatory scheme for new navigational hazards such those that arise from offshore wind developments.[34]

In conclusion, it was clearly evident from the debate in the Forum that EU member States have much to learn from Germany's mitigation strategy to cope with climate change including recent developments concerning offshore wind energy development in sea areas under national sovereignty and jurisdiction. One particularly impressive feature in this strategy is the steps taken by Germany to implement MSP and to resolve conflicting uses of the marine environment for different purposes at an early stage in the licensing process. Most notably, the broad scheme of public consultation in relation to the environmental impact of such projects is to be welcomed and demonstrates a real commitment to inclusive decision-making regarding the implementation of various regulatory measures governing offshore renewable energy development. In the final analysis and as a footnote to the decision of the International Court of Justice in the *North Sea Continental Shelf* cases, it is significant to note that it is offshore wind energy that makes the continental shelf and EEZ an important practical resource for Germany.

[34] Nov. 1984/suppl. 1987.

PART SIX

CITIES' CONTRIBUTIONS TO ENVIRONMENTAL PROTECTION

INTRODUCTORY REMARKS ON OPPORTUNITIES OF THE CITIES IN THE FIELD OF LOCAL CLIMATE CHANGE GOVERNANCE

Joachim Sanden[1]

1. Introduction

Governance as a political concept is already well-known.[2] Nevertheless, local climate change governance meets a comprehensive challenge. The C40 Cities Climate Leadership Group has stated (C40 Cities Climate Leadership Group 2011), 'cities are responsible for consuming two thirds of the world's energy and generating over 70% of its greenhouse gas emissions. This concentration of population, emissions and local, municipal oversight present a clear opportunity for impactful climate change policy'. According to the International Energy Agency (IEA), the urban population in Europe accounts for 69% of (European) energy use and thus the lion's share of greenhouse gas emissions in Europe.[3] This is the first reason beside others that cities matter in the context of facing climate change.[4] We require an intelligent burden sharing to meet the international goals, which may be renewed in Durban/South Africa (COP 17 of UNFCCC) after the expiration of the Kyoto Protocol in 2012. Local action has mirrored the global efforts, e.g. the Global Cities Convenant on Climate and the Carbon Cities Climate Registry parallel to the Cancun Agreements 2010. The 2011 Bonn Declaration of Mayors (2nd World Mayors Council on Climate Change) demands such as for 'building additional local institutional capacity to prepare, structure and manage large scale redevelopment'.[5] Because the Copenhagen climate negotiations failed, the idea for the

---

[1] Prof. Dr. Dr. Joachim Sanden is Head of Directorate in the Hamburg State Ministry for Urban Development and Environment, and Extraordinary Professor at Leuphana University Lueneburg.

[2] Hal K. Colebatch, 'Governance as a conceptual development in the analysis of policy' (2009) 3 Critical Policy Studies, 58–67, <http://dx.doi.org/10.1080/19460170903158107> accessed 07 July 2012.

[3] International Energy Agency (ed), World Energy Outlook 2008 (International Energy Agency, 2008).

[4] David Satterthwaite, *Shaping Urban Environment: Cities Matter. The World Bank Urban Strategy* (The World Bank 2009).

[5] Cf. <http://resilient-cities.iclei.org> accessed 07 July 2012.

local governments is to build their own processes and to integrate it to the upcoming global architecture. Therefore, cities in Europe have zoned down the big international binding or non-binding targets with emphasis on mitigation (that means reducing the sources or enhancing the sinks of greenhouse gases). The Covenant of Mayors in Europe concluded the aim to reduce carbon dioxide emissions by 20% by 2020 compared to 1990 levels. The U.S. Conference of Mayors' Climate Protection Agreement (USMCPA) made similar conclusions in 2005.[6] Besides, there are other significant agreements, including the Local Action 21 formed as a result of the First Earth Summit in Rio de Janeiro in 1992; and the Mexico City Pact of the World Mayors Summit on Climate in Mexico City 2010.[7] Substantial elements of the Mexico City Pact, signed by more than 140 cities, are to 'develop local adaption strategies to address the local impact of climate change', to 'register our emission inventories, commitments, climate mitigation and adaption measure and actions in a measurable, reportable and verifiable (MRV) manner' and to 'seek the creation of mechanisms that allow direct access to international funding for local climate actions'. Furthermore, there are the World Port Climate Conference (WPCC); the Ecoports Foundation;[8] the Green Capital Initiative of the EU Commission;[9] the C40 Cities Climate Leadership Group;[10] and the International Council for Local Environmental Initiatives (ICLEI).[11] Different cities have already passed climate change mitigation plans with greenhouse gas emission baselines.[12] For instance, Hamburg has passed ambitious climate protection goals of reducing its CO_2 emissions by 40% by 2020 and by 80% by 2050. On the other hand, the resources of the municipalities are rather limited. They are dependent upon local government capacities according to the state's constitution. Limited financial resources[13] (World Bank 2010) and probably restraints due to deregulation strategies on the national level

6 US Conference of Mayors (ed.), *Climate Protection Agreement—US Conference of Mayors*, <http://usmayors.org/climateprotection/agreement.htm> accessed 07 July 2012.

7 Cf. <http://www.wmsc2010.org/the-mexico-city-pact/> accessed 07 July 2012.

8 Cf. <http://www.ecoports.com/> accessed 07 July 2012.

9 Cf. <http://ec.europa.eu/environment/europeangreencapital/index_en.htm> accessed 07 July 2012.

10 Cf. <http://live.c40cities.org> accessed 07 July 2012.

11 Cf. <http://www.iclei.org/index.php?id=798> accessed 07 July 2012.

12 Christopher Kennedy, Anu Ramaswami, Sebastian Carney and Shobhakar Dhakal, *Greenhouse Gas Emission Baselines for Global Cities and Metropolitan Regions*. Proceedings of the 5th Urban Research Symposium Marseille, France: June 28–30, 2009.

13 World Bank (ed), *A City-wide Approach to Carbon Finance* (The World Bank 2010).

could be barriers. Verified local climate action with enhanced access to global climate funds is the trend of international efforts for Durban 2011.

2. Drivers for and Functions of the Cities in International Climate Protection

As a starting point after empowering and resourcing local governments we have to recognize that adaption processes are often locally driven. They are not as a rule 'top-down', but 'bottom-up'.[14] For example, the urban risk-analysis (e.g. risks of floods in low elevation coastal zones) should be better done on a local level better than on a national one.[15] The analysis gives the municipalities the chance and the justification to locate their drivers and to analyse their potential according to their main function in the process.

Dhakal[16] identified the following items as main drivers for the Characterization of Local Emissions Contexts: 'Compactness of the urban settlement, Urban zoning and functions, Nature of the transportation system, Income level and lifestyle, Energy efficiency of key technologies, Nature of economic activities, Building technologies and building floor space use, Waste management and climate factors'. We have to ask if climate law on a local level is also a crucial factor in this process.

Main functions of the big cities in international climate protection differs.[17] Cities could be actors themselves in the urban environment (public building, public procurement). Local policy makers governing the action of private actors (building codes, land use regulations). What we need is the implementation of an integrated approach, for example involving urban infrastructure providers. The city governments' roles in

---

14 Miranda A. Schreurs, 'From the Bottom Up. Local and Subnational Climate Change Politics' (2008) 17 The Journal of Environment and Development, 343–355; Nicholas Lutsey and Daniel Sperling, 'America's bottom-up climate change mitigation policy' (2008) 36 Energy Policy, 673–685.

15 World Bank (ed), *Understanding Urban Risk: A Methodology for Assessing Disaster and Climate Risk in Cities* (The World Bank 2010).

16 Shobhakar Dhakal, *Urban Energy Use and Greenhouse Gas Emissions in Asian Megacities*: Policies for a Sustainable Future (Institute for Global Environmental Strategies 2004).

17 Leonardo Meeus and Erik Delarue, *Mobilizing Cities towards a Low Carbon Future: Tambourines, Carrots and Sticks*, EUI Working Papers RCAS 2011/04 (Robert Schuman Centre for Advanced Studies, Florence, School of Regulation 2011).

climate change are described in detail by Alber and Kern:[18] 'A city government can act as one or more of the following:

- *Consumer*, intervening directly on municipal energy and transport consumption;
- *Planner and regulator*, orientating urban development and using authoritative powers to set mandatory conditions related to energy efficiency;
- *Provider and supplier*, investing in infrastructure in the transport, waste, and energy supply sectors, either directly or by owning companies providing such public services;
- *Enabler and adviser*, influencing other actors through information campaigns on sustainable behaviours or supporting them directly with incentives and counselling aimed at enhancing measures that can contribute to climate change mitigation.'

Combining actors and functions, Meeus and Delarue[19] have developed a concept of city smartness. They define first level city smartness by focussing on the actor's 'leading by example'. The second level is characterized by 'governing the private urban actors' by policy makers. The third level city smartness is according to the authors an 'integrated approach'. In my opinion, this concept does not reflect the appropriate realistic ranking of the steps of smartness. Therefore, I propose an earlier integration, involving own or even controlled urban infrastructure providers. Meanwhile, we should start to govern local policy makers steering the action of private actors. At least we should support the implementation of emission reduction plans and measures, which today are for the most part implemented at the national and regional levels (more emphasis on multi-level-governance).[20]

---

[18] Gotelind Alber and Kristine Kern, *Governing Climate Change in Cities: Modes of Urban Climate Governance in Multi-level Systems*, in Documentation Competitive Cities and Climate Change Conference, Milan 9–10 October 2008; Edoardo Croci, Sabrina Melandri, and Tania Molteni, *Comparing Mitigation Policies in Five Large Cities: London, New York City, Milan, Mexico City, and Bangkok*, in: Daniel Hoornweg, Mila Freire, Marcus J. Lee et al. (eds), *Cities and Climate Change—Responding to an Urgent Agenda*, ed. by the International Bank for Reconstruction and Development and The World Bank (The World Bank 2011) 55, 71.

[19] Cf. Meeus and Delarue (n. 17) 2.

[20] Joyeeta Gupta, 'The multi-level governance challenge of climate change' (2009) 4 Journal of Integrative Environmental Sciences, 131–137.

Because of these discussions, we find in numerous local climate plans similar main components, as the International Council for Local Environmental Initiatives—Local Governments for Sustainability (ICLEI)[21]—has summarized:

- 'The local Business-as-usual (BAU) scenario: Which assumptions and drivers have been considered in projecting future local emissions?
- The choice of the base year and of reduction targets: Which criteria do the local government follow in choosing and defining its reduction commitment?
- Mitigation measures: How relevant is each measure, and which roles does the local government play in each sector?
- Implementation and monitoring: Does the plan identify who will be responsible for the plan's implementation and the monitoring system?
- Financing: Does the plan addresses the funding of measures?'

3. Framing the Process of Transition to a Low Carbon Future

Political science uses the methodology of 'framing' to describe strategic planning. It addresses the role of the political issue definitions in the policymaking process and therefore analyses especially the political agenda-setting.[22] 'The climate change issue must be translated or framed to enable actors to work with this problem in a local context'.[23] The approach of 'framing', creating understandable categories for the abstract problem of (global) climate change and community building, is closely linked to understanding whether the governance process (regulation and integrated elements of co-steering) overlaps with central steering with

[21] Edoardo Croci, Sabrina Melandri, and Tania Molteni, *A Comparative Analysis of global City Policies in Climate Change Mitigation*, IEFE Working Paper (Bocconi University Working Paper No. 32, 2010).

[22] Falk Daviter, 'Policy framing in the European Union' (2007) 14 Journal of European Public Policy, 654–666; Sarah B. Pralle, 'Agenda-setting and climate change' (2009) 18 Environmental Politics, 781–799, <http://dx.doi.org/10.1080/09644010903157115> accessed 07 July 2012; Hyung Sam Park, Xinsheng Liu, Arnold Vedlitz, *Framing Climate Policy Debates: Science, Network, and U.S. Congress, 1976–2007*, Conference Proceedings 2010 (Southern Illinois University Carbondale 2010) <http://opensiuc.lib.siu.edu/cgi/viewcontent.cgi?article=1041&context=pnconfs_2010> accessed 07 July 2012.

[23] Gard Lindseth, 'The Cities for Climate Protection Campaign (CCPC) and the Framing of Local Climate Policy' (2004) 9 Local Environment, 325, 327.

hybrids.[24] Jänicke and Jörgens[25] differentiated between goal-and result-oriented management, environmental policy integration in the source sectors, cooperative governance in the narrow sense (i.e. as co-regulation) and participation of social actors. This means that lawyers should take cooperative and network governance[26] more and more into account. It is useful to look at how political science and new institutional economics explain—from the point of view of another discipline—the integration of three typical governance structures (institutional steering) and the contribution of the law. They start with the market (setting the legal preconditions for investing, ruling economic incentives). The second step is governing by hierarchy (obliging responsible parties to act) and at least by network (polycentric acting of the administration, and new coordination mechanisms).

As a next step, we have to ask for the consequences for the law. Is the dominance of weak governance mechanisms a reason for why there is no more space for legal action? Legal science has some difficulties to clarify the role of law in particular with regard to guaranteeing mechanisms for 'soft' forms of governance (or coordination). Layers still discuss, if the new instruments should act alternatively or complementary. Nevertheless, after a closer look, recent EU legislation such as REACH and the IED indicates a successful combination of new governance and law. Often, law is the basis of steering opportunities. Climate policy goes beyond current legal requirements. Opting-out is the way to gain the goals. Therefore, we can come to the result of co-existence between legislation and instruments in the new governance policies.[27] Soft law complements hard law.[28]

[24] David M. Trubek, Patrick Cottrell and Mark Nance, '"Soft Law", "Hard Law", and European Integration: Toward a Theory of Hybridity', Jean Monnet Working Paper 2/05 (New York School of Law 2005).

[25] Martin Jänicke and Helge Jörgens, 'Neue Steuerungskonzepte in der Umweltpolitik' (2004) 27 Zeitschrift für Umweltpolitik & Umweltrecht, 297–348.

[26] Erik-Hans Klijn, 'Networks and Governance: a perspective on public policy and public administration', in Ari Salminen (ed), *Governing Networks* (IOS Press 2003) 29–38.

[27] David M. Trubek and Louise G. Trubek, *The Coexistence of New Governance and Legal Regulation: Complementary or Rivalry?*, Paper Presented at the Annual Meeting of the Research Committee on the Sociology of Law, Paris, July 2005, <http://www.reds.msh-paris.fr/communication/docs/trubek.pdf> accessed 07 July 2012; Joanne Scott and Jane Holder, *Law and New Environmental Governance in the European Union*, NewGov Paper 2006, <http://www.ucl.ac.uk/laws/clge/docs/Scott%20and%20Holder.pdf> accessed 07 July 2012.

[28] Cf. Trubek, Cottrell and Nance (n. 24).

Of course, multi-level-governance characterizes the governance process[29]—international conventions, national legislation and local legal action. 'The relevance of national climate action planning for local responses is apparent' as Heinrichs, Aggarwal and Barton[30] wrote: 'From the perspective of city representatives, national plans are prepared primarily as a response to the international negotiations and processes of the United Nations Framework Convention on Climate Change with a primary focus on mitigation (instead of adaptation). More important, for the most part, they take a sectoral perspective and, with the exception of reference to coastal areas, have a nonurban bias. In some instances, the national plan has been "victimized" by conflicting political interests with contradicting and vague results. These arguments raise concerns about the adequacy and legitimacy of national plans in terms of guidance and support (...).'

After we have presented the methods or improved the concept, we can look at the disincentives and for those reasons behind barriers for the cities. According to Meeus and Delarue,[31] the first disincentives are market failures like the cost structure of energy supply etc. It seems difficult to influence the local actors to save money upfront rather than to save on future energy bills. Institutional failures are reflected in the slogans 'Not in my term/not my business', resulting in a lack of integrated solutions, for example in urban planning and transport. Third are multi-agent failures and in particular the task of extremely complex coordination of few and diverse actors (e.g. inclusive controlled infrastructure companies).

Regarding the major key enabling factors of the municipalities in local climate protection, there are two extreme positions: A more formal approach, represented e.g. by Bulkeley, Schroeder and Janda et al.,[32]

---

29 Simona Piattoni, 'Multi-level Governance: a Historical and Conceptual Analysis' (2009) 31 Journal of European Integration, 163–180 <http://dx.doi.org/10.1080/07036330802642755> accessed 07 July 2012.

30 Dirk Heinrichs, Rimjhim Aggarwal, Jonathan Barton et al., 'Adapting Cities to Climate Change: Opportunities and Constraints', in Daniel Hoornweg, Mila Freire, Marcus J. Lee et al. (eds), *Cities and Climate Change—Responding to an Urgent Agenda*, ed. by the International Bank for Reconstruction and Development and The World Bank (The World Bank 2010) 193, 221.

31 Cf. Meeus and Delarue (n. 17) 4–5.

32 Harriet Bulkeley, Heike Schroeder, Kathryn Janda et al., 'The Role of Institutions, Governance and Planning for Mitigation and Adaptation by Cities', in Daniel Hoornweg, Mila Freire, Marcus J. Lee et al. (eds), Cities and Climate Change—Responding to an Urgent Agenda, ed. by the International Bank for Reconstruction and Development and The World Bank (The World Bank 2010), 68–88.

requires that factors largely fall under the area of institutions and governance, which suggests that efforts to strengthen institutional and governance capacities in general at the city level would have the co-benefit of enabling better responses for adaptation. In contrast, Heinrichs, Aggarwal and Barton et al.[33] stated a more material understanding. They believe in the relevance of 'several factors that also emerge across the wider collection of studies, including the availability of information, the need for higher levels of awareness, synergies with existing priorities and programs, the existence of strong leadership, availability of dedicated resources, and adequate technical and financial capacities'. In my opinion, both approaches are correct, so we should combine them.

The potential answer of the municipal law becomes clear when we distinguish between three levels of motivation to move the citizens. This model is useful because it seems unnecessary to use a more complex system of motivators.[34] As Meeus and Delarue[35] argue, as 'tambourines' to raise awareness, we already find conclusions of local bodies, e.g. the Covenant of Mayors and the International Council for Local Environmental Initiatives (ICLEI) etc. The second category is the 'carrots', which means the way to enable action. Items especially include the framing of a subsidy concept for the city as a public actor to compensate for higher costs of energy efficiency projects (Local Energy Finance Scheme). Another step can be to create legal guidance for Public-Private-Partnership Projects (e.g. re-buy of district heating systems; combined heat and power—CHP).[36] Finally, yet importantly, we can mention as an example the Hamburg′ Eco-Partnership with industry aimed at motivating businesses to participate on a voluntary basis. The third category is obliging action, so we call them 'sticks'. This means climate protection legislation, for example for the Energy Performance of Buildings (by specified energetic standards and technical regulations; or procurement (contracts award manual). Maybe traffic regulations may also be included.

Local legislation can and should use the framework created by national law and international law. The extent to which a municipality is able to

---

[33] Cf. Heinrichs, Aggarwal, Barton et al. (n. 29) 9.

[34] Maike Sippel and Till Jenssen, 'What About Local Climate Governance? A Review of Promise and Problems', Working Paper, November 2009, 7, available at SSRN: <http://ssrn.com/abstract=1514334> accessed 07 July 2012.

[35] Cf. Meus and Delarue (n. 17) 6–7.

[36] Jochen Monstadt, 'Urban Governance and the Transition of Energy Systems: Institutional Change and Shifting Energy and Climate Policies in Berlin' (2007) 31 International Journal of Urban and Regional Research, 326–343.

implement local climate policies also depends on 'a city's competences and authority to regulate climate-relevant issue areas'.[37] In some countries, national 'Global Warming Law' sets an expectation for local climate legislation, but it is questionable whether the domestic law does really provide the required legal competences.[38]

To give an example for local legal action, I should focus on the Hamburg Climate Protection Act (HCPA), which is a successful example how we can set legally binding targets by including local governments. HCPA provides the basis for the setting of energetic regulations within the scope of urban land use planning. Hamburg is thus a national trendsetter as regards the obligatory use of sustainable energy within the scope of land development plans. In accordance with the HCPA, the new installation of electric storage heating systems is prohibited. In the years 2000 to 2006, the number of such heating systems was thus already reduced by 25%. Furthermore, Hamburg endeavours to enforce by law the replacement of all remaining systems. In December 2007, Hamburg passed a Hamburg Climate Protection Ordinance, which specifies the highest level of heat protection requirements for buildings in all of Germany. Hamburg plans to revise this Climate Protection Ordinance now in order to establish the passive house standard. HCPA contains a special cost-efficiency benchmark for energy saving measures in public buildings such as programmes for lamps, boilers and refrigerator replacement.

4. Key Design Elements of Further Local Legal Action

The next chapter describes some upcoming problems with the design of regulation of such CPAs. Municipal actors should decide whether they set up voluntary guidelines or better mandatory regulation beyond national standards. They should explore the framework of self-governance in the field of climate protection with its two main aspects of mitigation or adaption to climatic challenges. The chance to regulate public procurement

[37] C40 Cities Climate Leadership Group (C40) (ed), *Letter from C40 Sao Paulo Summit to Rio + 20*, United Nations Conference on Sustainable Development Statement of C40 Steering Committee on the role of cities in tackling climate change, 02.06.2011, <http://live.c40cities.org/storage/resources/FINAL%20Rio20%20Communique.pdf> accessed 07 July 2012.

[38] Noriko Sugiyama and Tsuneo Takeuchi, 'Local Policies for Climate Change in Japan' (2008) 17 The Journal of Environment and Development, 424, 425 and 428; Sippel and Jenssen (n. 34).

in a better way should be used. The legal shape of the planning of urban infrastructure is quite a substantial question, so too is the need to clear the legal preconditions for municipal energy/heat supply and many other items.

It is apparent that in many countries a deregulation strategy on the national level would provide opportunities. But it could also restrain local (legal) activities. Options are the cooperation with business, use of eco-audits & enforcement relief for audited sites and the degradation of standards and regulations (e.g. conversion of existing standards to set rules (especially in waste law)). Restraints are apparent. Often there is less scope for the municipalities because of the high degree of regulation, growing task allocations to municipalities and decreasing financial allocations. We recognize the retention of material environmental legal standards and ensuring functional equivalence—that means the required conditions for any deregulation. The legislature has broad discretion concerning constitutional and supranational legitimacy, but it has to pay attention to legal requirements (e.g. the EU-IPPC/IE-Directive). Specific attention lies on the principle of local self-government: Communities should secure the interest of an effective administration of the local affairs. They take the protection and the exercise of their remit to ensure municipal responsibility (general welfare and security) into account. They believe in a ban on full, unrestricted alienation of communal tasks and therefore in the necessity of influence and control rights.

We should underline the high importance of local participation and democratic legitimacy[39] when we draft local legislation to contribute to climate change activities. Local democratic legitimacy and municipal welfare are more important than competition considerations. Some scholars focus upon the local warranty liability for actions on a local level. The partnership with local boards and co-governance brings additional influence channels. The key to securing the fulfilment of this complex task could be quite easy if we look to the well-known UN-ECE-Aarhus Convention dealing with three pillars: information, participation & access to justice. Although the scope of application of the Aarhus Convention with 44 Parties from Europe and Central Asia is not yet worldwide, their spirit is

39 Ank Michels and Laurenk de Graaf, 'Examining Citizen Participation: Local Participatory Policy Making and Democracy' (2010) 36 Local Government Studies, 477–491, <http://dx.doi.org/10.1080/03003930.2010.494101> accessed 07 July 2012.

widely accepted. On this basis, we can justify institutionalized citizen participation (e.g. in the usage of renewable energy sources (RES)) and thus improvement of environmental protection and energy saving. The second item is the improved coordination of individual needs and municipal tasks by better information (monitoring and reports), better integration of stakeholders and better communication. Consequently, we see the advantages of a better identification and reduction of conflict potential with private interest groups and superordinate authorities (political resistance, blocking legal procedures etc.) and the basis for voluntary engagement.

5. Conclusions

Municipal actors have recognized their responsibilities and resources to give a substantial contribution to the national action plans for climate change multi-level-governance. Based on their understanding of drivers and municipal functions they start the framing of local policy to meet the challenges of adaption, as well as mitigation. Local law can play a vital complementary function in the process of the transition to a low carbon future; there is co-existence between legislation and other instruments. Raising awareness and finding the way to enable action are the main issues. Key elements are the regulation of local energy finance schemes or the creation of legal guidance for public-private-partnership projects in the district heat and power sector. Finally yet importantly, we should regulate obliging legal action (e.g. for the energy performance of buildings, procurement and some sectors) more. This local legislation could face the framework of national deregulation strategies, which may exist. At the very least, the recognition of the basic principles of the UN-ECE Aarhus Convention, in concrete information, participation and access to justice, reflects the high importance of local democratic legitimacy and participation when we draft local legislation to contribute to climate change activities.

CONCLUSIONS OF FORUM II: CITIES' CONTRIBUTIONS TO ENVIRONMENTAL PROTECTION

Martin Huber[1] and Joachim Sanden[2]

1. Introduction

The Free and Hanseatic City of Hamburg as hosting municipality is the 2011 European Green Capital[3]—Hamburg will utilize the year to send an impulse throughout Europe, disseminating its best practices in the built urban environment, nature and climate. Therefore, it was the wish of the Board of the Hamburg International Law Conference to offer a specific platform for the contribution of municipal governments. In FORUM II, which was called 'Cities' Contribution to Environmental Protection', approximately thirty participants from all over the world discussed in two sessions the chances of cities to give contributions to climate protection activities. After some introductory remarks by the Board of this Forum, nine speakers (see the complete list below) gave their personal impression of what to do in the future. At the end of the vital discussion the participants agreed to publish a paper of common understanding, which is as follows:

2. Importance of Cities for Environmental Protection

Hamburg International Law Conference (HIELC), FORUM II, recognizes that half the world's population lives in urban areas and that a significant portion of the human activities that lead to global climate change are concentrated in cities.

---

[1] Martin Huber was Head of Directorate in the Hamburg State Ministry for Urban Development and Environment, now he is Executive Director in the Hamburg State Ministry of Economic Affairs, Traffic and Innovation.

[2] Prof. Dr. Dr. Joachim Sanden is Head of Directorate in the Hamburg State Ministry for Urban Development and Environment, and Extraordinary Professor at Leuphana University Lueneburg.

[3] <http://hamburggreencapital.eu/> accessed 07 July 2012.

FORUM II confirms that municipal governments have considerable authority, for instance over land-use planning, waste management, transportation issues and energy consumption and development (e.g. energy efficiency, solar panel roofs), all of which have implications for greenhouse gas (GHG) emissions.

FORUM II agrees that a city government can act as a public actor and last but not least as an enabler and adviser for the private sector.

3. Present City Action

FORUM II welcomes Cities Climate Protection (CCP) sponsored by the International Council for Local Environmental Initiatives (ICLEI), as well as other initiatives all other the world, such as the World Mayors Summit on Climate in Mexico City on 21.11.2010, the Covenant of Mayors in Europe, the U.S. Conference of Mayors' Climate Protection Agreement (USMCPA) in 2005, Local Action 21, World Port Climate Conference (WPCC); Ecoports Foundation; Green Capital Initiative; C40 Cities Climate Leadership Group and others.

The recognition of climate change as a local concern is a crucial issue. Commitments to control local GHG emissions need compliance. Responsibilities of local governments should be especially focused upon:

- Climate change policy by local governments;
- Local Action Plans;
- Appointment of Climate Change Action Officers to promote activities to mitigate global warming;
- Regional and Local Centers to promote and enhance global warming prevention activities;
- Establishment of Regional and Local Partnership Council;
- Local Energy Finance Schemes;
- Climate protection legislation, for example for the Energy Performance of Buildings (by specified energetic standards and technical regulations); procurement; public transportation systems.

FORUM II believes in the necessity for development and implementation of policies and programs to reduce emissions. Municipalities are now responding to the need to adapt to the long lasting change in weather patterns generated by climate change—climate-research and adaptation to adverse effects of climate change play a pivotal role in responding to recent and future risks.

Cities are frequently far more ambitious than their national governments with regard to GHG reduction (e.g. Chicago, Osaka, and Hamburg). Most cities participating in schemes such as CCP are 'pioneers'. FORUM II wants to encourage them. For example, several cities in Africa, Asia and other continents already use the gain of methane gas from landfill sites as a vital important contribution to reduce GHG emissions.

4. Problems and Answers

Climate change is a global phenomenon. When the linkage of policy response is more obvious (e.g. controlling GHG emissions leads to a higher air quality), reframing climate change as a local issue may be more likely.

FORUM II wants to put its special focus on resources and capacities. Adaption and mitigation processes are often locally driven, not 'top-down'. On the other hand, regional and local governments depend on national or even international regulations and political commitments. But the main issue is that city actions depend on local government capacities according to the nation state′s constitution (Multi-Level-Governance) and limited financial resources.

FORUM II recognizes similar challenges, for example:

- supply with drinking water/waste water sewage and treatment;
- waste disposal/landfill sites/gaining energy from waste by combustions or methane production;
- transportation;
- investment allocation;
- reducing CO_2-emissions;
- (potential) power supply shortages; and
- rising energy demand.

FORUM II realizes a broad range of municipal answers, including:

- Restructuring industry and creating a new transportation system (*Shanghai*);
- Improving energy efficiency of different industrial branches, buildings, private households, etc. (*Shanghai, Osaka, Chicago, Hamburg*);
- Improvements in public transportation and use (*Auckland, Hamburg*);
- Integrated waste recycling concepts (*Nashik, Dar es Saalam*);
- Landfill /waste water gas management (*Nashik, Dar es Salaam*);

- Regional and local spatial planning respecting climate objectives (*Nashik, Osaka, Auckland, Hamburg*);
- Improve net metering standard and enhanced market structures and regulatory framework (*Chicago, Hamburg*);
- Maintenance of metropolitan urban limits (*Auckland*);
- Local climate action plan/local ordinances, including energy efficiency measures and install local greenhouse trading systems, labeling systems for products, and green building programs (*Osaka*);
- Reinforcing legal enforcement and supervision (*Shanghai*);
- Improvement of procurement, for example in renewable energies (*Chicago, Hamburg*); and
- Economic incentives (*Shanghai, Chicago, Hamburg*).

FORUM II believes that climate regulations on a local level can play an important role to set the legal framework for cities' actions in the field of CCP.

In the field of adaption, FORUM II realizes several local activities reducing the adverse impact of climate change, like the identification of vulnerabilities and risks including priorities, the drafting of adaptation actions and their implementation. Issues that are specifically relevant to local governments are the public risk perceptions and the development of new standards to improve or upgrade the robustness of future infrastructure.

FORUM II states that there could and should be potential answers of the local regulations. The task is framing a legal concept for the city as a public actor. Local governments can (of course) set tougher regulations (e.g. emissions standards) than national standards, when the national standards are regarded as minimum standards and when the local ordinance does not disturb the object of other national laws.

FORUM II underlines the strong function of local democratic legitimacy and participation. The idea is to integrate business, labor and politics. FORUM II pays attention to local democratic legitimacy and municipal welfare and agrees to the local warranty liability. FORUM II requires strengthening partnership with local boards and co-governance. It is in line with the idea of the UN-ECE-Aarhus Convention with the key words information, participation & access to justice. FORUM II asks for an institutionalized citizen participation, for example in the usage of renewable energy sources (RES) and thus improvement of environmental protection and energy saving, the improved coordination of individual needs and municipal tasks by better information (monitoring and reports), better integration of stakeholders and last but not least better communication.

FORUM II believes that there is a focus on the communication strategy to offer practical every day climate protection solutions for our citizens. It is vital to place the citizens at the centre of discussion. This includes increasing awareness in all things related to climate change and environmental protection.

5. Outlook on Future Possibilities for Cities to Act on Climate Protection

Participants in FORUM II were encouraging:

- municipal leaders to think and act locally as well as globally when it comes to climate protection knowing that financial consequences have to be faced, and
- cities to learn from each other and jointly develop solutions for shared or similar problems.

Participants in FORUM II stated that from the point of view of international environmental law legal action on the local level will have an enhancing effect on the cities actions. The goal is conversing policy with reality. Answers of the local climate law can be the setting up of climate protection legislation, for example for the energy performance of buildings, improving procurement to optimize CCP and structuring general public transport. Adaption means especially integration of this issue into local building and/or water legislation. In general, local adaptation strategies could facilitate communication needs ensuring a broad stakeholder involvement. FORUM II believes that we can get more efficient permitting procedures. Further regulation can be useful when framing a subsidy concept for the city as a public actor to compensate for higher costs of energy efficiency projects (Local Energy Finance Scheme) and some other measures. The legal shape of the planning of urban infrastructure as well the creation of legal preconditions for municipal energy supply will be other challenges.

Participants in FORUM II were optimistic that the cities will find their path between voluntary guidelines and mandatory regulation in line with and beyond national standards. On the basis of self-governance the cities will use the framework for action.

Speakers in Forum II:

- Prof. Dr. Kenneth Palmer, University of Auckland, New Zealand
- Prof. Howard Learner, Executive Director Environmental Law & Policy Center, Chicago/IL Chicago, U.S.A.
- Richard M. Katiti, Dump Site Manager, Dar es Salaam City Council, Dar es Salaam, Tanzania
- Dr. Benno Hain, Climate Coordinator of the Free and Hanseatic City of Hamburg, Hamburg, Germany
- R.K. Pawar, Head of Environmental Cell, City of Nashik, India
- Prof. Dr. Noriko Okubo, Osaka State University, Osaka, Japan
- Wang Qiang, Deputy Director, Shanghai Solid Waste Management Center, Shanghai, PR of China
- Martin Huber, Head of the Legal Office of the Ministry of Urban Development and Environment, Hamburg
- Prof. Dr. Dr. Joachim Sanden, Head of Directorate Soil/Contaminated sites of the Free and Hanseatic City of Hamburg

PART SEVEN

RESULTS

RESULTS: CLIMATE CHANGE

Hans-Joachim Koch[1]

Ladies and Gentleman,

we had in the field of climate protection policy and law four excellent presentations, in the working groups important statements and lively, sometimes sharp discussions. What are the results?

May I remind you of the questions I put forward in my opening speech. These were the questions

- of a legally binding warming cap (2°C aim),
- of legally binding emissions reduction aims,
- of finding a fair burden sharing among the industrialized states as well as between industrialized and developing states,
- of a new instrumental diversity in climate protection in addition to ETS, and
- of addressing the emissions of cars, ships, airplanes and agriculture more effectively.

With respect to these questions I would like to sum up as follows:

1. Nearly all participants agreed on the necessity of legally binding decisions on the international level. Some of them have doubts about the chance to obtain such an agreement. In the opinion of Lavanya Rajamani, India would be able to agree, if there was a binding agreement on a fair burden sharing. Jacqueline Peel has stressed that legally binding aims on the international level would support and improve Australian climate protection policy and law heavily. Many agreed that this is true also for other laggered states.
2. In both working groups there was a consensus that national and regional initiatives in addition to international conventions are necessary. It is this level where instrumental diversity especially comes into play. But—as Marc Pallemaerts correctly insisted—on the international

[1] Prof. Dr. Hans-Joachim Koch, Hamburg University, Germany.

level in climate regulation there already exists a lot of diversity with respect to the duties of the convention's parties.

3. Finally, the debate on the 'top-down' and the 'bottom-up' approaches was in my opinion negatively influenced by the ambiguity and vagueness of these terms. I think a rational climate policy needs as justification for regulation a top-down approach as follows:

- On the top there is the warming cap (2°C aim).
- With respect to this target the next step is to limit the concentration of GHG in the atmosphere.
- On this basis decisions about the global emissions reduction aims are needed.
- The next step has to be a decision on a fair burden sharing.

Finally there are decisions on the adequate instrumental diversity.

RESULTS: ENVIRONMENTAL HAZARDS RELATED TO SHIPPING

Doris König[1]

On Friday, 16 September 2011, the conference focused on two sections, namely (1) the reduction of greenhouse gases (GHG) from shipping, and (2) the prevention of shipping accidents.

With regard to the reduction of GHG from shipping, there is a conflict between two legal forums, the global forum in the IMO and the regional forum in the EU, because the EU is not satisfied with the outcome of the activities in the IMO and has threatened to take its own regional measures. So far the IMO has agreed upon technical and operational measures which fall within its core competence. Both the IMO and the EU agree that this is only a first step, and that in order to reach the 2°C goal more effective measures are necessary.

Working Group 1 (GHG reduction) discussed the prospects for more effective measures on both levels. The outcome was as follows:

In the IMO:

- There will be ongoing discussions on market-based measures (MBMs).

These discussions are difficult, because MBMs touch upon politically sensitive issues. Some countries do not want to accept global measures in the IMO forum in order not to compromise their negotiating position in the UNFCCC forum.

- A comprehensive package could be negotiated by 2015/16.
- In order to get universal acceptance, it is certain that these measures must generate funds for mitigation and adaptation measures in developing countries.

The use of income generated from the reduction of GHG from shipping for investments in developing countries was regarded as a matter of justice/fairness.

---

[1] Prof. Dr. Doris König, Bucerius Law School, Hamburg.

In the EU:

- The Commission will probably table a proposal in 2012 which will come into effect 3 to 4 years later.
- It is not yet decided which measures shall be taken.
- The integration of shipping into the EU Emissions Trading System is just one option; other options are under discussion as well.
- EU regulation is regarded as a 'bottom-up approach' if other regions follow suit with similar regulations.

Finally, it has been suggested that the IMO and the EU could share the work: IMO should deal with technical and operational measures, whereas the EU and possibly others should adopt complementary market-based measures.

In Working Group 2 (prevention of shipping accidents), the discussion focused on the relationship between global and regional EU measures. The outcome was as follows:

On the global level:

- Almost all risks that could cause shipping accidents are covered by international treaties adopted by the IMO and others; there are hardly any gaps left.
- Therefore, in the future the human factor has to be taken into account more seriously.
- Working conditions of seafarers and social considerations should come more into focus in order to prevent accidents and to make the seafarer's profession more attractive.
- New forms of training are needed for seafarers to help them maintain high safety and security standards.

It was agreed that the adoption of such measures is not only a task for the IMO, but should also be pursued by other international and regional organizations, such as the ILO and the EU.

On the EU level:

- It was generally agreed that EU/regional regulations are important and supplementary to IMO regulations.

- In most cases the EU fully transposes IMO rules and standards into EU law and, thus, makes sure that these provisions are implemented by the Member States.
- In some cases, however, the EU explores the limits of international law set by the UN Convention on the Law of the Sea and the IMO or other treaties, by adopting stricter or more burdensome measures.
- Whereas in some cases, such as the phasing-out of single hull tankers, the IMO adapts its more lenient global to stricter EU standards, in other cases the EU approach might backfire.
- The EU should, therefore, carefully deliberate whether non-compliance with international law is the right option to choose.

LIST OF PARTICIPANTS

Adam Ahen

Christina Albrecht; *German Offshore Wind Energy Foundation*

Haseeb Anwar Ul Haq; *Society for Enforcement of Rule of Law, Islamabad, Pakistan*

Prof. Dr. Ivo Appel; *Universität Augsburg, Germany*

Ra Aida Ardabilli

Sascha Arnold

Rosalía Arteaga; *Fundación Fidal, Quito, Ecuador*

Susann Barahona; *Botschaft von Mexiko*

Lawrence D. Barchue; *Member State Audit and Internal Oversight Services, IMO, London, England*

Laura Bärthel; *Hamburg, Germany*

Carsten Bartholl; *Taylor Wessing, Hamburg, Germany*

Tim Bäuerle; *Rechtsanwälte Zimmermann Gretz Trautmann Bäuerle, Heidelberg, Germany*

Dr. Tariq Bilal; *Society for Enforcement of Rule of Law, Islamabad, Pakistan*

Prof. Dr. Daniel Bodansky; *School of Sustainability, Arizona State University, USA*

Dagmar Bork; *Behörde für Stadtentwicklung und Umwelt, Hamburg, Germany*

Hildur Bragadóttir; *Environment Agency of Iceland, Reykjavík, Iceland*

Monika Breuch-Moritz; *Federal Maritime and Hydrographic Agency (BSH), Hamburg, Germany*

Dr. Manuel Cadmus; *Hamburg, Germany*

Dr. Jorge Caillaux; *Sociedad Peruana de Derecho Ambiental (SPDA), Lima, Peru*

Richard Calland; *University of Capetown, South Africa*

Pablo Cardenal

Friedrich Catoir; *Ex MFA GER, Clamecy, France*

Hong Chang

Michael Chartrey; *Bucerius Law School, Hamburg, Germany*

Prof. Dr. Aldo Chircop; *Dalhousie University, Halifax, Nova Scotia, Canada*

Silke Marie Christiansen LL.M.; *Berlin, Germany*

Andreas Chrysostomou; *MEPC, IMO*

Klaus de Buhr; *Behörde für Stadtentwicklung und Umwelt, Hamburg, Germany*

Jan de Haan; *Dr. Weiland & Partner, Hamburg, Germany*

Alice de Juvigny; *Bucerius Law School, Hamburg, Germany*

Robert Derksen; *Swiss Climate, Hamburg, Germany*

Ana Di Pangracio; *Fundación Ambiente y Recursos Naturales (FARN), Buenos Aires, Argentina*

Simon Djallo Patchoke; *Ministry of the Environment and Protection of Nature, Yaoundé, Cameroon*

Pamela Domröse; *Research Centre for European Environmental Law, University of Bremen, Germany*

Regina Döller

Anja Eikermann; *Georg-August-University Göttingen, Germany*

Uta Elsner

Dan Engoren; *Bucerius Law School, University of Bremen, Germany*

Prof. Dr. Astrid Epiney; *Université de Fribourg, Switzerland*

Hans-Juergen Ermisch; *Klemm & Partner Rechtsanwälte, Hamburg, Germany*

Dr. Frank Evers; *White & Case LLP, Hamburg, Germany*

Ioannis Fakidarakis; *Universität Hamburg, Germany*

Eoin Fannon; *Office of the Attorney General, Dublin, Ireland*

Glóey Finnsdóttir; *Ministry for the Environment, Reykjavík, Iceland*

Jochen Flasbarth; *Umweltbundesamt, Dessau, Germany*

Prof. Dr. Heidi Foth; *Sachverständigenrat für Umweltfragen (SRU), Germany*

Leonhardt Frank; *Leonhardt & Blumberg, Hamburg, Germany*

Dr. Claudio Franzius; *Universität Hamburg, Germany*

Hans Gabányi; *Behörde für Stadtentwicklung und Umwelt, Hamburg, Germany*

Jana Gassner; *Hamburg, Germany*

Anne-Christin Gläß; *Universität Leipzig, Institut für Umwelt- und Planungsrecht, Germany*

Karsten Gold; *Airbus Operations GmbH, Hamburg, Germany*

Phillip Goltermann; *Drees + Sommer, Hamburg, Germany*

Gwendoline Gonsaeles; *MDK, Brussels, Belgium*

Barbara Görnandt

Dr. Joachim Greinert; *Stadtreinigung Hamburg, Germany*

Kerstin Gröhn; *Universität Hamburg, Forschungsstelle Umweltrecht, Germany*

Lena Groth; *Bucerius Law School, Hamburg, Germany*

Markus Groth

Michael Günther; *Rechtsanwälte Günther, Hamburg, Germany*

Martin Hack; *Rechtsanwälte Günther, Hamburg, Germany*

Caroline Hagenberg; *Dabelstein & Passehl, Hamburg, Germany*

Dr. Sabine Hain; *Behörde für Wirtschaft, Verkehr und Innovation, Hamburg, Germany*

Dr. Benno Hain; *Behörde für Stadtentwicklung und Umwelt, Hamburg, Germany*

Christopher Hanson; *Bucerius Law School, Hamburg, Germany*

Rashid Haroonur; *Society for Enforcement of Rule of Law, Islamabad, Pakistan*

Prof. Dr. Christian Heitsch; *Brunel University London, England*

Dr. Anja Hentschel; *University of Kassel, Germany*

Christoph Heßling; *Bucerius Law School, Hamburg, Germany*

Jacqueline Hilbert; *Federal Environment Agency (Umweltbundesamt), Dessau, Germany*

Dr. Matthias Hoes; *Behörde für Stadtentwicklung und Umwelt, Hamburg, Germany*

Dr. Timo Hohmuth; *Grenius Rechtsanwälte, Hamburg, Germany*

Karin Horn

Katherine Houghton; *Universität Trier, Germany*

Martin Huber; *Behörde für Stadtentwicklung und Umwelt, Hamburg, Germany*

Theresa Ilgner; *Bucerius Law School, Hamburg, Germany*

Sabine Juli

Patricia Jonason; *Södertörn University, Stockholm, Sweden*

Dr. Christian Kahle; *White & Case LLP, Hamburg, Germany*

Shohreh Kananizaeh; *Nooretouba University, Tehran, Iran*

Dr. Jan Kappet; *VG Hamburg, Germany*

Dr. Jörn Kassow; *White & Case LLP, Hamburg, Germany*

Richard M. Katiti; *Dar es Salaam City Council, Tanzania*

Elminaz Khatami; *Landgericht Frankfurt, Germany*

Carolin Kiess; *BFN, Germany*

Kathrin Kim; *Universität Hamburg, Germany*

Dr. Bernd Kisilowski; *Airbus Operations GmbH, Hamburg, Germany*

Prof. Dr. Hans-Joachim Koch; *Universität Hamburg, Forschungsstelle Umweltrecht, Germany*

LL.M. Christina Koenig; *Hamburg Port Authority, Germany*

Prof. Dr. Doris König; *Bucerius Law School, Hamburg, Germany*

Dr. Holger Kraft; *CMS Hasche Sigle, Hamburg, Germany*

Dr. Berend Krüger; *IAG—Ihlenberger Abfallentsorgungsgesellschaft mbH, Germany*

Jörg Kuhbier; *Kuhbier Rechtsanwälte, Hamburg, Germany*

Fiona Kunihira; *Kampala, Uganda*

Mari-Liis Kupri; *Ministry of the Environment, Tallinn, Estonia*

Holger Lange; *Behörde für Stadtentwicklung und Umwelt, Hamburg, Germany*

Prof. Dr. Mojib Latif; *Leibniz-Institut fuer Meereswissenschaften, Kiel, Germany*

Braet Laurie; *LDR Advocatenkantoor, Gent, Belgium*

Prof. Dr. Howard A. Learner; *Environmental Law & Policy Center Chicago, IL, USA*

Prof. Dr. Dieter Lenzen; *Universität Hamburg, Germany*

Yun Li; *Shanghai Academy of Environmental Sciences, P. R. China*

Erika Lindhorn

Heiki Lindpere; *Estonian Maritime Academy, Tallinn, Estonia*

Nicolai Livonius; *Airbus Operations GmbH, Hamburg, Germany*

Dr. Ronán Long; *School of Law, National University of Ireland Galway*

Jianting Lu; *Shanghai Academy of Environmental Sciences, P. R. China*

Christian Maaß

Gabriele Mahro; *Behörde für Stadtentwicklung und Umwelt, Hamburg, Germany*

Michael Mainusch; *Behörde für Stadtentwicklung und Umwelt, Hamburg, Germany*

Mark Major; *European Commission, DG Climate Action*

María Máñez Costa; *Climate Service Center, Hamburg, Germany*

Melanie Manns; *Leuphana University Lüneburg, Germany*

Wenpei Mao; *Shanghai Environmental Protection Bureau, P. R. China*

Ralf Marxen

Dr. Friederike Mechel; *Behörde für Stadtentwicklung und Umwelt, Hamburg, Germany*

Friedrich Joachim Mehmel; *Rechtsstandort Hamburg e.V., Hamburg, Germany*

Björn Meilen; *Bucerius Law School, Hamburg, Germany*

Moritz Meister; *Leuphana Universität Lüneburg, Germany*

Prof. Dr. Meinhard Hilf; *Bucerius Law School, Hamburg, Germany*

Wulf Menzel; *Universität Hamburg, Germany*

Prof. Dr. Dirk Messner; *Deutsches Institut für Entwicklungspolitik (DIE), Germany*

Christin Mielke; *HafenCity Universität Hamburg, Germany*

Manfred Milde-Büttcher; *Vattenfall Europe Business Services GmbH, Hamburg, Germany*

Dr. Peter C. Mohr; *Mohr Rechtsanwälte, Hamburg, Germany*

Dr. Ulf-Henning Möker; *Verwaltungsgericht Hamburg, Germany*

Bernd Möller

Lutz Morgenstern

Subashini Murugesan; *Consulate General of India, Hamburg, Germany*

Tina Mutert; *Federal Environment Agency Germany (UBA), Dessau, Germany*

Euphrasie Francoise Ngono; *Ministry of Urban Development and Environment, Yaoundé, Cameroon*

Dr. Nico Nolte; *Federal Maritime and Hydrographic Agency (BSH), Hamburg, Germany*

Prof. Dr. Sebastian Oberthür; *Institute for European Studies, Vrije Universiteit Brussel, Belgium*

Sveinung Oftedal; *Norwegian Ministry of the Environment*

Masanori Okada; *Waseda University Institute of Comparative Law, Japan*

Prof. Dr. Noriko Okubo; *Osaka State University, Japan*

Uta Ordemann; *Verband Deutscher Reeder e.V., Hamburg, Germany*

Heinz Otto; *Bundesverband Windenergie e.V., Landesverband Hamburg, Germany*

Philip Pahl; *Bucerius Law School, Hamburg, Germany*

Prof. Dr. Marc Pallemaerts; *ACELS, University of Amsterdam, The Netherlands*

Prof. Dr. Kenneth Palmer; *University of Auckland, New Zealand*

Martin Paus; *Universität Bonn, Hamburg, Germany*

Ramesh Kashinath; *Pawar Nashik, India*

Dr. Jacqueline Peel; *Melbourne Law School, Australia*

Michael Peper; *Behörde für Stadtentwicklung und Umwelt, Hamburg, Germany*

Alexander Pirang; *Bucerius Law School, Hamburg, Germany*

Matthias Plötzke; *Verband Deutscher Reeder e.V., Hamburg, Germany*

Dr. Ursula Prall; *Kuhbier Rechtsanwälte, Hamburg, Germany*

Dr. Matthias Pflughaupt

Prof. Dr. Alexander Proelß; *Universität Trier, Germany*

Wang Qiang; *Shanghai Solid Waste Management Center, P. R. China*

Prof. Dr. Lavanya Rajamani; *Centre for Policy Research, New Delhi, India*

Prof. Dr. Ulrich Ramsauer; *Universität Hamburg, Germany*

Ulrich Rausch; *Behörde für Stadtentwicklung und Umwelt, Hamburg, Germany*

Prof. Dr. Catherine Redgwell; *UCL Faculty of Laws, England*

Dr. Moritz Reese; *Helmholtz Centre for Environmental Research, Leipzig, Germany*

David Reichwein; *Max-Planck-Institute for Comparative Public and International Law, Heidelberg, Germany*

Dr. Judith Reuter; *Behörde für Stadtentwicklung und Umwelt, Hamburg, Germany*

Raja Saifur Rehmann; *Services Law Associates*

Martin Richter; *Hamburg, Germany*

Dr. Henrik Ringbom; *EMSA*

Prof. Dr. Michael Rodi; *IKEM e.V:, Berlin, Germany*

Katrin Roth; *Rechtsanwälte Günther, Hamburg, Germany*

Muhammad Saifullah; *Society for Enforcement of Rule of Law, Islamabad, Pakistan*

Dr. Pierre Sames; Germanischer Lloyd SE, *Hamburg, Germany*

Prof. Dr. Dr. Joachim Sanden; *Ministry of Urban Development and Environment, Free and Hanseatic City of Hamburg, Germany*

Annalisa Savaresi; *University of Copenhagen, Faculty of Law, Denmark*

Olaf Schaefer; *Alfred-Wegener-Institut, Germany*

Dr. Michael Schäfer; *Freshfields-BruckhausDeringer LLP, Hamburg, Germany*

Prof. Dr. Arndt Schmehl; *Universität Hamburg, Germany*

Gert-Jürgen Scholz; *Bundesministerium für Verkehr, Bau und Stadtentwicklung, Bonn, Germany*

Stefan Schoeneck; *Landwirtschaftsministerium Mecklenburg-Vorpommern, Germany*

Dr. Bernadette Seehafer; *Kreditanstalt für Wiederaufbau, Frankfurt, Germany*

Verena Seemann

Stefanie Seliger; *Hamburg, Germany*

Anke Seyfried; *Springer-Verlag GmbH, Heidelberg, Germany*

Amber Sharick; *European Renewable Energies Federation, Brussels, Belgium*

Akiho Shibata; *Kobe University, Japan*

Dr. Ruediger Siechau; *Stadtreinigung Hamburg, Germany*

Torben Skaanild; *BIMCO*

Dr. Ronald Steiling; *Graf von Westphalen, Hamburg, Germany*

Mirja Storck; *Universität Hamburg, Germany*

Lutz Strack

Svenja Stropahl; *Leuphana Universität Lüneburg, Germany*

Dr. Bernd Thyssen; *Hamburg, Germany*

Martin Triipan; *Raidla Lejins & Norcous, Tallinn, Estonia*

Dr. Kai Trümpler; *Federal Maritime and Hydrographic Agency (BSH), Hamburg, Germany*

Julian Udich; *Bucerius Law School, Hamburg, Germany*

Masakuni Ueta; *Kobe University, Japan*

Dr. Roda Verheyen; *Rechtsanwälte Günther, Hamburg, Germany*

Alexandra von Saldern; *Gesellschaft für Internationale Zusammenarbeit, Germany*

Sophie van Boemmel

Sarah Wack; *Hamburg, Germany*

Kersten Wagner-Cardenal; *White & Case LLP, Hamburg, Germany*

Tilo Wallrabenstein; *Verband Deutscher Reeder e. V., Hamburg, Germany*

Anne-Barbara Walter; *Federal Environment Agency (Umweltbundesamt), Dessau, Germany*

Qiang Wang; *Shanghai Solid Waste Management Center, P. R. China*

Matthias Weckerling; *Bundesministerium der Justiz, Germany*

Jun Wei; *Shanghai Environmental Protection Bureau, P. R. China*

Lilly Weidemann; *Berlin, Germany*

Katja Weiler; *IZES gGmbH, Saarbrücken, Germany*

Daniel Welss; *Universität Hamburg, Forschungsstelle Umweltrecht, Germany*

Jennifer Wesche; *Behörde für Stadtentwicklung und Umwelt, Hamburg, Germany*

Martin Wickel; *HafenCity University Hamburg, Germany*

Zhao Wie; *Shanghai Environmental Protection Bureau, P. R. China*

Nicolai Woelki; *Deutscher Nautischer Verein von 1868 e.V., Germany*

Dr. Sarah Wolf; *Dabelstein & Passehl, Hamburg, Germany*

Dr. Nina Wolff

Dr. Andreas Wolowski

Frank Wolke; *Deutsche Emissionshandelsstelle, Germany*

Dr. Ulrich Wollenteit; *Rechtsanwälte Günther, Hamburg, Germany*

Teppei Yagyu; *Waseda University School of Law, Japan*

Farzaneh Zahed; *University of Hamburg, Germany*

Cathrin Zengerling; *Rechtsanwälte Günther, Hamburg, Germany*

Angela Zhao; *Bucerius Law School, Hamburg, Germany*